TERTIARY LEVEL BIOLOGY

The Behavioural Ecology of Ants

JOHN H. SUDD
University of Hull
and
NIGEL R. FRANKS
University of Bath

Blackie

Glasgow and London

Published in the USA by
Chapman and Hall
New York

Blackie & Son Limited,
Bishopbriggs, Glasgow G64 2NZ
7 Leicester Place, London WC2H 7BP

Published in the USA by
Chapman and Hall
in association with Methuen, Inc.
29 West 35th Street, New York, NY 10001

© 1987 Blackie & Son Ltd
First published 1987

All rights reserved.
No part of this publication may be reproduced,
stored in a retrieval system, or transmitted,
in any form or by any means,
electronic, mechanical, recording or otherwise,
without prior permission of the Publishers.

British Library Cataloguing in Publication Data

Sudd, John H.
The behavioural ecology of ants.—
(Tertiary level biology).
1. Ants—Behavior 2. Insects—Behavior
3. Social behavior in animals
I. Title' II. Franks, Nigel R.
595.79'60451   QL568.F7

ISBN 0-216-92245-3
ISBN 0-216-92246-1 Pbk

Library of Congress Cataloging-in-Publication Data

Sudd, John H. (John Hilton)
The behavioural ecology of ants.

(Tertiary level biology)
Includes index.
1. Ants. I. Franks, Nigel R. II. Title.
III. Series.
QL568.F7S89 1987   595.79'6   87-6361
ISBN 0-412-01561-7 (Chapman & Hall)
ISBN 0-412-01571-4 (Chapman & Hall: pbk.)

Photosetting by Thomson Press (I) Ltd., New Delhi.
Printed in Great Britain by Bell & Bain Ltd.

The Behavioural Ecology of Ants

TERTIARY LEVEL BIOLOGY

A series covering selected areas of biology at advanced undergraduate level. While designed specifically for course options at this level within Universities and Polytechnics, the series will be of great value to specialists and research workers in other fields who require a knowledge of the essentials of a subject.

Recent titles in the series:

Locomotion of Animals	Alexander
Animal Energetics	Brafield and Llewellyn
Biology of Reptiles	Spellerberg
Biology of Fishes	Bone and Marshall
Mammal Ecology	Delany
Virology of Flowering Plants	Stevens
Evolutionary Principles	Calow
Saltmarsh Ecology	Long and Mason
Tropical Rain Forest Ecology	Mabberley
Avian Ecology	Perrins and Birkhead
The Lichen-Forming Fungi	Hawksworth and Hill
Plant Molecular Biology	Grierson and Covey
Social Behaviour in Mammals	Poole
Physiological Strategies in Avian Biology	Phillips, Butler and Sharp
An Introduction to Coastal Ecology	Boaden and Seed
Microbial Energetics	Dawes
Molecule, Nerve and Embryo	Ribchester
Nitrogen Fixation in Plants	Dixon and Wheeler
Genetics of Microbes (2nd edn.)	Bainbridge
Seabird Ecology	Furness and Monaghan
The Biochemistry of Energy Utilization in Plants	Dennis

Contents

INTRODUCTION viii

Chapter 1 SOCIAL BEHAVIOUR AS A SELFISH STRATEGY 1
  1.1 Kin selection 2
  1.2 Inclusive fitness and sex ratios 4
    1.2.1 Multiple queens 5
    1.2.2 Multiple mating 5
    1.2.3 Male production by workers 5
  1.3 Parental care and manipulation 7
  1.4 The evolution of polygyny 8
    1.4.1 Secondary polygyny 12
    1.4.2 Polygyny as a form of parasitism 17
    1.4.3. The behaviour of polygynous queens 18
    1.4.4 Oligogyny 19
  1.5 Dominance hierarchies in workers 19
  1.6 Cooperation and competition 22

Chapter 2 THE PHYLOGENY OF ANTS 24
  2.1 The origin of ants 24
    2.1.1 The ant family tree 24
    2.1.2 A Mesozoic fossil ant 26
    2.1.3 A living fossil ant 28
    2.1.4 Adaptation to liquid feeding 29
  2.2 The subfamilies of ants 31
    2.2.1 Myrmeciinae 31
    2.2.2 Ponerinae 32
    2.2.3 Dorylinae 32
    2.2.4 Pseudomyrmecinae 33
    2.2.5 Myrmicinae 33
    2.2.6 Dolichoderinae 37
    2.2.7 Formicinae 38
    2.2.8 Further reading 39

Chapter 3 ANT ECONOMICS 40
  3.1 Economies of scale 41
    3.1.1 The colony-founding stage 41
    3.1.2 The ergonomic stage 41

	3.1.3 The reproductive stage	43
3.2	Colony life-history strategies	43
	3.2.1 The schedule of growth, investment and reproduction	43
	3.2.2 Knowing when to split	46
3.3	The flow of resources within the colony	50
	3.3.1 Food exchange between workers	50
	3.3.2 The flow of food to larvae	51
	3.3.3 The course of food flow in the society	53
3.4	Nest construction	55
	3.4.1 Benefits of the nest structure	56
	3.4.2 Nest-excavation techniques	57
	3.4.3 Above-ground nest structures	58
	3.4.4 Import of special materials for nests	60
	3.4.5 Compounding special materials for nests	61

Chapter 4 WHO DOES WHAT, AND WHEN? 65

4.1 How ants are employed: how many tasks are performed in ant colonies? 66
4.2 Temporal polyethism: production lines based on an age-based division of labour 69
  4.2.1 Caste systems 71
  4.2.2 Task allocation 74
  4.2.3 Adaptive demography and caste efficiency 75
4.3 Conflicts over the division of labour 75
  4.3.1 A morphological division of labour in monomorphic ants 77
4.4 Physical castes 79
  4.4.1 Allometry as a developmental constraint on caste evolution 80
4.5 The economics of caste ratios 84
  4.5.1 Time and motion studies 87
  4.5.2 Case study: the ergonomics of leaf-cutter ants 87
  4.5.3 Case study: the ergonomics of foraging in army ants 91
4.6 Caste ratios and social homeostasis 94

Chapter 5 COMMUNICATION 98

5.1 Ant signals and language 98
5.2 Recognition of nestmates 100
5.3 Pheromonal communication 106
  5.3.1 Alarm pheromones 107
  5.3.2 Multiple pheromones 108
5.4 Communication in recruitment 111
  5.4.1 Simple cooperative hunting 112
  5.4.2 Group recruitment 113
  5.4.3 Mass recruitment 113
5.5 Sex pheromones 116
  5.5.1 Queen pheromones 116
  5.5.2 Other sexual pheromones 117

Chapter 6 ANTS AS PARTNERS 120

6.1 Ants in the ecological community 121
6.2 Ants and plants 121
  6.2.1 Ants and extrafloral nectaries 122
  6.2.2 Ants as allelopathic agents of trees 123
  6.2.3 Transport of seeds by ants 124

CONTENTS vii

	6.2.4 Ant gardens	125
	6.2.5 Ants and epiphytes	125
	6.3 Ants and other insects	126
	6.3.1 Ants and aphids	127
	6.3.2 Ants and Lepidoptera	131
	6.4 The cost-benefit balance in mutualism	133

Chapter 7 ANTS EXPLOITING ANTS 137
  7.1 Types of exploitation 137
    7.1.1 Mugger ants 137
    7.1.2 Claim-jumpers 138
    7.1.3 Thief ants 138
    7.1.4 Guest ants 139
  7.2 The temporary and permanent parasitic ants 140
    7.2.1 Infiltration by parasitic queens 140
    7.2.2 Temporary parasites 142
  7.3 The evolution of inquilines 143
    7.3.1 The ultimate cuckoo ants 146
  7.4 Slavery 149
    7.4.1 Amazon ants 149
    7.4.2 The tiny slave-makers 151
    7.4.3 Imprinting and slave-making 153
    7.4.4 The evolution of slave-making 154
    7.4.5 Are slave-makers degenerate? 158

Chapter 8 ANT ECOLOGY 161
  8.1 Competition 162
  8.2 Economics of territorial defence 168
  8.3 Foraging for the most profitable prey 176
  8.4 Ants as predators and prey: army ant foraging ecology 179

REFERENCES 188
INDEX 200

INTRODUCTION

This book is concerned with two problems: how eusociality, in which one individual forgoes reproduction to enhance the reproduction of a nestmate, could evolve under natural selection, and why it is found only in some insects—termites, ants and some bees and wasps. Although eusociality is apparently confined to insects, it has evolved a number of times in a single order of insects, the Hymenoptera. W. Hamilton's hypothesis, that the unusual haplodiploid mechanism of sex determination in the Hymenoptera singled this order out, still seems to have great explanatory power in the study of social ants. We believe that the direction, indeed confinement, of social altruism to close kin is the mainspring of social life in an ant colony, and the alternative explanatory schemes of, for example, parental manipulation, should rightly be seen to operate within a system based on the selective support of kin. To control the flow of resources within their colony all its members resort to manipulations of their nestmates: parental manipulation of offspring is only one facet of a complex web of manipulation, exploitation and competition for resources within the colony. The political intrigues extend outside the bounds of the colony, to insects and plants which have mutualistic relations with ants.

In eusociality some individuals (sterile workers) do not pass their genes to a new generation directly. Instead, they tend the offspring of a close relation (in the simplest case their mother). This reproductive relation (the queen) produces offspring some of which are in turn sterile. Her reproduction is in a sense delayed while resources are used to create a social structure with a generation overlap, both sterile and fertile members and the tending of non-offspring. The social structure is always closed and exclusive; that is, it is hostile not merely to non-kin, but to any members of its own species, kin or not, that do not bear its particular society marker or visa. This exclusiveness ensures that resources collected by a member are not usually put at the disposal of non-kin. As well as this well-defined social structure, eusocial insects always have a well-developed physical infrastructure. This means that instead of devoting their resources to producing new reproductive and sterile kin, they use a substantial amount of their resources of energy, time and materials in producing for example a nest or nests, and sometimes other structures like routes or tunnels outside the nest or shelters for their food resources. The use of material and energetic resources to produce structures is of course also found in non-social animals like solitary bees and wasps, caddis-flies and spiders. Subsocial insects, like aphids or locusts, on the other hand, devote all their resources to reproducing with the greatest possible speed.

We have tried to set these two topics, the social structure and the physical infrastructure of ant colonies, in the context of a system whose function is to confine the benefits of the society to kin. The system is not totally successful in this, and we extend our treatment to the exploitation of the society by enemies such as social parasites and mutualists.

Chapter 1 is concerned with the problems which the sociality of ants sets for evolutionary and genetic theory, in particular the question of how far the members of the colony adopt the Three Musketeers' approach of 'all for one and one for all' and how far they pursue their own genetic interests. Is the colony in fact united, or are there conflicts among its members? In Chapter 2 we look briefly at the capabilities which the individual ant brings to its social life, and to the exploitation of the natural resources of the colony's environment. We also survey briefly the diversity of environmental resource exploitation and of social organization in ants. Chapter 3 is concerned with the way in which an ant colony allocates these resources between the production of new reproductive males and females (which propagate the genes of some or all of its members into new colonies), and investment in biological infrastructure, like nurse workers, or in physical infrastructure such as a nest. Chapter 4 follows the way in which individual members of the colony differ in size, shape or behaviour and how this applies the colony's resources of time, energy and material efficiently. In Chapter 5 we look at the communication systems that confine resources within the colony and ensure the coordination of the colony's efforts.

The remaining chapters turn to the relations of the colony with other organisms; the way in which they exploit them or are exploited by them is the topic of Chapter 6. The special case of exploitative relations between ant species is considered in Chapter 7. Finally, Chapter 8 looks at the whole question of the ecology of ant colonies, their density, size and dispersal, in the light of the mechanisms we have described.

It is a pleasure to thank Mr R. Wheeler-Osman for redrawing the figures; he and Ms C. Ellis also provided some excellent photographs.

JS
NRF

CHAPTER ONE

SOCIAL BEHAVIOUR AS A SELFISH STRATEGY

There are perhaps 12 000 living species of ants, and their organization undoubtedly represents the pinnacle of social evolution in animals. For this reason ant colonies are referred to as being **eusocial**: literally, truly-social. This type of organization has three characteristics: (i) overlapping adult generations; (ii) cooperative brood care; (iii) more or less non-reproductive workers or helpers. Eusociality occurs in other insects only among the termites, and some wasps and bees.

Social insects in general, and ants in particular, have occupied centre stage in evolutionary biology over the last two decades because eusociality, and particularly the occurrence of a sterile work force, represents an evolutionary puzzle. If natural selection were to work through the direct inheritance of traits, sterility would disappear after one generation. Even the production of a proportion of sterile offspring would seem to be a grave disadvantage. For this reason the question of how the trait of sterility could evolve by natural selection was a source of special concern to Charles Darwin. The presence of sterile workers in ants represented, in Darwin's own words, 'one special difficulty, which at first appeared to me insuperable, and actually fatal to my whole theory'. The solution Darwin (1859) proposed was that sterile forms evolved because they are 'profitable to the community' and 'selection may be applied to the family, as well as to the individual'. In other words the selective advantage might be at the level of the colony, i.e. the family made up of the maternal queen and her daughter workers. Sterile workers could lead to the rapid growth and reproduction of colonies by helping in their parental nest, and the trait of worker sterility could be passed on through the extra reproductive offspring produced by these cohesive and super-productive family units.

Darwin's proposal was a remarkable insight, but it does not solve the problem of how the sterility of social insect workers arose in the first place.

The first 'workers' were almost certainly not sterile but merely stayed in the maternal nest where they helped their mother to raise more offspring. Sterility was probably a secondary trait once this colonial or family life had begun. So the question of how sterility arose must await a solution to the problem of how certain individuals might become helpers in the nest. But why should a female choose to stay in the parental nest rather than attempt to start her own highly productive family? In modern terms, how could genes that code for helping the reproduction of relatives become more abundant in future gene pools, if such genes also led to an individual producing fewer of its own direct descendants? Two solutions to this enigma have been put forward: kin selection and parental manipulation.

1.1 Kin selection

This hypothesis was proposed by Hamilton (1964). He suggested that under certain circumstances female Hymenoptera (which includes workers of ants, bees and wasps) could pass more copies of their genes into future generations by rearing sisters and brothers instead of their own sons and daughters. This possibility is in large part due to the unusual mechanism of sex determination in Hymenoptera. In this group females develop from fertilized and hence diploid eggs, with a set of chromosomes from their mother and from their father. Males are produced from unfertilized eggs and are therefore haploid, inheriting their single set of genes from their mother. All the sperm produced by a haploid male carry an identical set of genes. Thus female Hymenoptera with the same father and mother will have identical copies of genes in the half of their diploid genotype that they inherited from their father. This means that a very high proportion of the genes of hymenopteran sisters will be identical by common descent. On average, full sisters will have in common half of the genes from their mother and all the genes from their father. Since half of their total gene complement came from each parent, the average fraction of genes that sisters will have in common is given by $(\frac{1}{2} \times \frac{1}{2} + \frac{1}{2})$, i.e. $\frac{3}{4}$.

However, hymenopteran mothers share only $\frac{1}{2}$ of their genes with their daughters, and $\frac{1}{2}$ of their genes with their sons. So hymenopteran sisters are more closely related to each other than are mothers to their own offspring (Figure 1.1 and Table 1.1). Thus a gene which programs a hymenopteran female to raise sisters in her mother's nest, rather than daughters in her own independent nest, is likely to spread in the genotypes of these sisters. This scheme also appeared to solve the problem of why all the workers in hymenopteran societies are females.

SOCIAL BEHAVIOUR AS A SELFISH STRATEGY 3

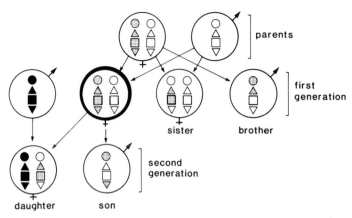

Figure 1.1 A schematic representation of genetic relationships in an ant family tree. A gene present in a mother's genome has a 50% chance of occurring in any one of her daughters or sons. However, since males do not have fathers and are haploid, all their sperm are alike. Therefore, full sisters have identical genes inherited from their father. This means that any gene present in a female has a 75% chance of being present in her full sister. Kin selection therefore favours the rearing of fertile sisters rather than daughters. By contrast a brother and sister, on average, will have only 25% of their genes in common by direct descent from their mother's genome. Males are therefore unlikely to evolve as kin-rearing workers. Adapted from J. Maynard Smith (1978) *Scientific American* **239**, 176–192.

Table 1.1 Degree of relatedness between either a male or female ant and their close relatives in an ant family tree.

	Mother	Father	Sister	Brother	Son	Daughter	Niece
Female	0.5	0.5	0.75	0.25	0.5	0.5	0.375
Male	1	0	0.5	0.5	0	1	0.25

By this simple argument Hamilton had apparently solved one of the greatest puzzles in evolution. Hamilton's idea caused a revolution in evolutionary biology, with its suggestion that traits could evolve that are detrimental to the survival or reproduction of one organism, if they are disproportionately advantageous to the survival and reproduction of the kin of that organism. This is the idea of *inclusive fitness*. An organism's fitness depends not only on its own survival and reproduction but on the fecundity of its relatives. The contribution of a relative's offspring to an individual's fitness will depend on the coefficient of relatedness (r) between that individual and its relatives' progeny. The coefficient of relatedness is the probability that a gene in one individual is an identical copy by descent of a gene in another individual. Such r values are calculated for an ant family tree in Table 1.1.

1.2 Inclusive fitness and sex ratios

However, without a further refinement Hamilton's argument will not fully explain the evolution of sterile female workers in Hymenoptera. The problem is that in ants the coefficient of relatedness between a sister and her brother is only $\frac{1}{4}$. Thus if a female stays with her mother and helps to raise equal numbers of brothers or sisters, or invests energy equally in them, her average coefficient of relatedness to the siblings she produces will be only $(\frac{1}{4} + \frac{3}{4})/2$, i.e. $\frac{1}{2}$. Such a helping female would have done equally well by investing equally in the two sexes of her own offspring, as she would have a coefficient of relatedness of $\frac{1}{2}$ with both her sons and her daughters.

Trivers and Hare (1976) showed that the sterile workers' inclusive fitness would best be served by a 3:1 ratio of investment in favour of female sexual offspring rather than males, whereas the queen's optimum investment ratio would be 1:1. Trivers and Hare predicted that if Hamilton's theory explained the evolution of workers, then ant colonies should produce a 3:1 ratio of investment in favour of sexual females rather than males. They presented data, mostly from ants, that seemed to support their prediction. Even more ingeniously Trivers and Hare (1976) suggested that parasitic ants, such as slave-makers (see Chapter 7), provided an additional test for their theory. Queens of parasitic ants use workers of other species to raise their broods. Since the inclusive fitness of such slaves is unaffected whatever the sex-ratio of the parasites they raise, selection on the parasitic queen to produce a 1:1 sex ratio will not be resisted by her slaves. Trivers and Hare's (1976) seminal paper has been criticized on a number of grounds. First, the data they used to test their hypotheses is of rather mixed quality because it was derived mostly from studies that were not specifically investigations of sexual investment by social insects. Second, their statistical treatment of these data was arguably invalid. (But see Nonancs (1986) who recently presented a new and more rigorous analysis of sex ratios in ants which also supports the predictions of Trivers and Hare (1976).) Thirdly, the assumptions underlying their predictions have also been challenged.

First, workers within a colony will be $\frac{3}{4}$ related only if the colony consists of a single mother queen who has been inseminated only once (i.e. all workers are full sisters), and second, the prediction that they would favour a 3:1 investment ratio in favour of females assumes that workers do not contribute to the production of males by laying their own haploid eggs. We now consider each of these assumptions in turn.

1.2.1 Multiple queens

Many social insect colonies have more than one functional queen, and such polygyny is particularly common in ants. Even if these queens are a mother and her daughters or a set of sisters, the workers they produce will have on average less than $\frac{3}{4}$ of their genes in common, and such workers will be even less closely related on average to the sexual progeny that their polygynous colony produces. Consider the case in which workers raise offspring of a queen who is their sister: the worker's relatedness to such offspring will be only $\frac{3}{8}$ for both nephews and nieces (Table 1.1). Nevertheless, the corresponding relatedness for raising nephews and nieces in diploid organisms would be even less ($r = \frac{1}{4}$). In this way haplodiploidy still favours this form of kin cooperation and kin selection. In the case of workers helping sisters there should be no conflicts over sex ratio, as both the queen and her sister workers should favour a 1:1 investment in the reproductive sexuals.

1.2.2 Multiple mating

If a queen mates with several males then the daughters she produces need not have the same father, and hence degrees of relatedness within the nest will be progressively reduced with each additional insemination that the queen has received. Nevertheless, for a given number of matings by the queen, relatedness between sisters is higher in Hymenoptera than in diploid species. As Cole (1983) has shown, if workers are only about 10% more efficient at producing reproductives within a eusocial colony than they would be solitarily, then two matings by the queen will still not override the selective advantage to the workers of being eusocial. Furthermore, multiple inseminations may be a secondary trait that is largely confined to the most advanced species, such as army ants (see Chapters 2 and 8). Thus multiple mating by the queen may have arisen only after ants or their ancestors had become eusocial, as an adaptation to very large colony size—where a queen needs a large reserve of sperm to produce a huge diploid work-force over a long lifetime.

1.2.3 Male production by workers

Because hymenopteran males are produced from unfertilized eggs, females that have not been inseminated can produce sons by parthenogenesis (virgin birth). In many species of ants, workers do have ovaries and contribute many of their own sons to the male production of their colony (e.g. in *Myrmica*; Smeeton, 1981). Workers are more closely related to their own sons than to their brothers and so should prefer to rear their own sons rather than those of the queen. Similarly the queen should prefer her own sons to those of her daughter workers. Thus production of males is another possible source of conflict within ant colonies. Most intriguingly, if workers produce all the males in a colony a 1:1 sex ratio is expected, just as would be the case if the queen was fully in control. Thus conclusions about who is winning in conflicts within a colony should not be based solely on the sex ratio it produces. We will return to worker parentage of males in section 1.2.3.

These problems associated with multiple queens, multiple mating and worker reproduction, which would reduce average relatedness in ant colonies, have cast doubt on Trivers and Hare's (1976) simple and elegant analysis. Clearly it is important to determine coefficients of relatedness not just in hypothetical ant colonies but in real ones.

Gel electrophoresis has made it possible to determine the enzymatic phenotypes of individuals, deduce genotypes from these and estimate what proportion of their genes are likely to be in common by immediate descent.

For example, individuals with heterozygotic alleles at a certain site on their chromosomes will produce slightly different forms of the same enzymes (isozymes) in their tissues. When these tissues are crushed into a buffer which is then run onto a gel plate and subjected to an electrical gradient, the enzymes will migrate to different zones where they can be marked and recognized with specific stains. Heterozygotes produce two stained bands, one for each isozyme, whereas homozygotes produce just one distinct band. Many types of isozyme can be recognized and used to make statistical estimates of gene frequencies in different sets of individuals.

Using such electrophoretic techniques, nestmate workers have been shown to have indeed approximately 75% of their genes in common, by direct descent, in a number of ant species whose colonies are characterized by single, once-mated queens. This confirms Hamilton's (1964) predictions. The species in which such $\frac{3}{4}$ coefficients of relatedness have been demonstrated include, for example, *Solenopsis invicta* and certain *Rhytidoponera* species (Ross and Fletcher, 1985; Ward, 1983). The results of comparative electrophoretic studies will be discussed in more detail in section 1.4.

For these reasons, though both the data and many assumptions of Trivers and Hare's (1976) elaboration of Hamilton's (1964) theory have been criticized it remains clear that haplodiploid sex determination and the asymmetric degrees of relatedness it confers are important in hymenopteran eusociality. This assertion is supported by the fact that eusociality has evolved independently 12 times in insects; 11 times in hymenoptera and only once elsewhere, in the termites (Andersson, 1984).

1.3 Parental care and manipulation

Haplodiploidy by itself does not mean that eusociality is likely to evolve. As Andersson (1984) has pointed out, many mites, ticks, thrips, whiteflies and scale insects, and some beetles, are also haplodiploid; yet none of these groups has eusocial species.

The simple point is that for eusociality to evolve, helping behaviour has not only to be a possibility, but has to have significant quantitative advantages. For a hymenopteran female to gain by giving up her own reproduction in favour of helping her mother, she must be able to contribute significantly to her mother's fitness, and hence her own inclusive fitness. If she can raise 1.5 times more surviving offspring on her own than she is able to add siblings to her mother's brood, then selection will favour her personal reproduction (Brockman, 1984). For this reason, helping behaviour can most easily be understood as an option in a species which already has parental care in a form in which non-parents can also profitably contribute their efforts. In insects this implies that there will be a distinct and recognizable nest in which a parent or parents tend, feed and defend their brood (Eickwort, 1981). This has led to the suggestion (Andersson, 1984) that the defensive stings and manipulative mandibles of female Hymenoptera were important factors in the repeated evolution of eusociality in this group. Once other members of the family become helpers in the nest, economies of scale can contribute to the efficiency of the family-community through a division of labour in which certain individuals specialize as, say, nest builders, defenders, brood nurses or foragers (see Chapter 4). In other words, helpers at the nest must be able to contribute significantly to the 'infrastructure' of their family's economy.

Because parental care seems to be such an important precursor of eusocial evolution, some authors (see Andersson, 1984 for a review), have suggested that the prime factor in the evolution of workers was parental manipulation. For example, if a mother were to restrict the food she gave to certain of her daughters they might develop into such small individuals that they would be unable successfully to found a nest alone. Their only means of increasing their (inclusive) fitness would be to stay in the parental nest and contribute to its economy. However, the conflicts of interest which lead to parental manipulation and the idea of kin 'voluntarily' helping as a result of kin selection are far from being mutually exclusive. Indeed, the close relatedness between the manipulated daughters and the siblings that they help to raise reduces the selection pressure to escape from manipulation, so kin selection and parental manipulation were probably both intimately involved in the evolution of eusociality.

Because all living ants are fully eusocial we can only guess about many of the stages in their evolution. Nevertheless both theories, kin selection and parental manipulation, imply that ant societies are not simply productive communities in which all the individuals are striving for the same goals; on the contrary, social life is rife with conflicting interests, manipulative and counter-manipulative. We now consider two important variations on this theme of conflicts of interest within ant societies—polygyny, the coexistence of more than one queen in a nest, and dominance hierarchies, conflicts between certain related workers.

1.4 The evolution of polygyny

Although colonies with single queens, which have mated only once, have extremely closely related workers this cannot be the case in the many ant colonies that have multiple queens. For example, about half of the European ant species are classified as polygynous (Buschinger, 1974*b*). This poses interesting questions about the relatedness of the queens to their workforce and vice versa, and raises the possibility of conflicts of interest within polygynous colonies. Before we can explore the selection pressures that have given rise to polygynous colonies we should first recognize that polygyny in ants has many distinct forms.

Firstly, though a colony may possess many morphological queens they may not all contribute equally to the growth of the colony or the production of sexuals. This inequality of the queens can arise because (i) some of the queens are not inseminated and therefore cannot produce female offspring; (ii) they are different ages and have different degrees of ovarian development which may translate into different rates of egg laying; (iii) some of the queens are behaviourally or chemically inhibited from egg-laying by other queens in the nest.

These points have been illustrated by Buschinger (1968) who made a thorough survey of monogyny and polygyny in 11 species of European *Leptothorax*. He determined the status of queens by dissections in which the presence or absence of sperm in the spermatheca (the sack in which females store sperm after insemination) and ovarian activity were noted. In *Leptothorax acervorum*, *L. muscorum*, *L. tuberum* and also in the parasite *L. kutteri* (see Chapter 7) for example, some colonies are polygynous and others monogynous. These species can be described as 'facultatively polygynous'. In some of the polygynous colonies only a few of the queens were fertilized and egg-laying, whilst the others were either young queens, just fertilized and not yet laying, or unfertilized queens, young or old. Unfertilized queens were never found to have mature oocytes.

Colonies of *L. gredleri*, by contrast, often contained several wingless females but there was always only one egg-layer. Additional fertilized females might be present, but the function of their ovaries was suppressed by some unknown mechanism. For this reason Buschinger (1968) described *L. gredleri* as being functionally monogynous. The other six species of *Leptothorax* he examined were exclusively monogynous in the field. In *L. unifasciatus*, for example, obligatory monogyny is guaranteed by a special behavioural mechanism. If two or more fertile queens of this species are placed in a cage from which they cannot escape but which gives access to workers, and they are then presented to a colony of queenless workers, all but one of the queens will be killed by the workers. When two fertile females are caged together without workers, they fight each other to the death.

Thus even within one genus of ants there are many different forms of polygyny in addition to monogyny. Such findings have led Buschinger (1974*b*) to suggest, in common with a number of other authors, that monogyny is typically the primitive condition in ants and polygyny is usually a secondary trait.

Polygyny can arise in two different ways within the life of a colony. When a colony is founded by a single queen this is referred to as *haplometrosis*; such colonies can become secondarily polygynous later in their life. Second, a colony may be founded by *pleometrosis*, i.e. several queens come together to initiate a new colony. These two types of polygyny are fundamentally different. Indeed, pleometrosis is believed to lead only very rarely to permanent polygyny in ants (Hölldobler and Wilson, 1977). After the colony has begun to grow, the extra queens are killed or driven out of the nest, or else the colony splits with each queen taking part of the workforce.

Recently Bartz and Hölldobler (1982) have analysed the costs and benefits, for the individuals concerned, of pleometrotic associations in the honey-pot ant, *Myrmecocystus mimicus*, which lives in the deserts of the southwestern United States. Established colonies of these honey-pot ants are always monogynous, and their nests are regularly spaced out because they destroy small colonies that try to nest too close to them. New colonies of this species, however, are often initiated by 2–9 queens. Laboratory experiments indicate that new colonies raid one another, stealing brood which develops into workers that then join in further raiding. In this way

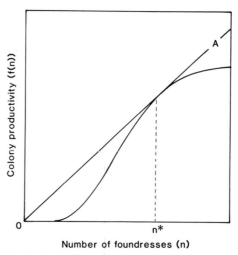

Figure 1.2 Graphical solution to the problem of optimal foundress group size. The straight line AO has a slope of $f(n)/n$, the ratio that selection should act to maximize. The maximum value of $f(n)/n$ is given by the point at which line AO is tangential to $f(n)$. n^* is therefore the optimal foundress group size. (After Bartz and Hölldobler, 1982.)

certain colonies grow much more quickly than others, and to their direct detriment. Nests with more queens initially grow more quickly and are most likely to win in these battles. However, individual queens in groups have a lower chance of surviving, because the workers eventually discard or kill all but one of the queens. For these reasons there should be an optimum group size that will give an individual queen the best chance of taking over a nest that will itself survive. Bartz and Hölldobler incorporated these factors into a model to predict the optimum queen number in an association. Their solution is shown in Figure 1.2. The optimum group size is the one that yields the greatest colony productivity, i.e. workers or pupae produced, per queen. In the laboratory, Bartz and Hölldobler (1982) found that the foundress group size maximizing the number of pupae or workers per queen was between 2 and 3 (Figure 1.3). This coincides quite closely with the most common group sizes in nature which are between 2 and 4. Thus queens are probably selfishly joining the best size of group that will give each of them the best overall chance of survival.

The workers that kill all but one of the queens might be killing their own mother and reducing their own inclusive fitness, since there is no evidence that they recognize their mother. In this case, how could selection favour the evolution of such behaviour? If workers could recognize only the

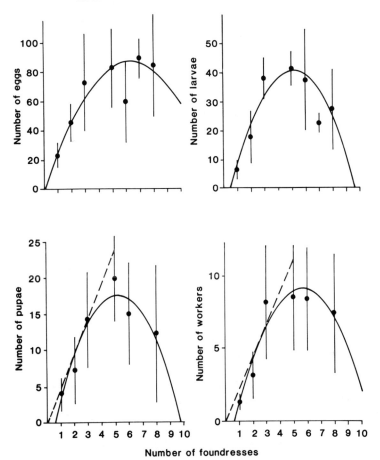

Figure 1.3 Data for colony production as a function of the number of foundress females ($X \pm$ S.D.) in incipient colonies of the honey ant, *Myrmecocystus mimicus*, in the laboratory. The curves were fitted using the method of least mean squares. (a) Maximum number of eggs observed in incipient nests. The best fit regression is $y = -0.819 + 28.4x - 2.62x^2$; $r = 0.469$: $p < 0.001$:48df. (b) Maximum number of larvae observed: $y = -12.9 + 21.2x - 2.06x^2$; $r = 0.633$: $p < 0.001$:32df. (c) Maximum number of pupae observed. $y = 5.23 - 8.82x - 0.849x^2$; $r = 0.364$: $p < 0.001$:28df. (d) Number of workers in each colony on 3 September 1975, approximately 7 weeks after the beginning of the experiment. $y = -2.88 + 4.28x - 0.377x^2$; $r = 0.431$: $p < 0.001$:23df.

The optimal foundress group size can be determined from these graphs using the method outlined in Figure 1.2. The foundress group size that maximizes the number of pupae or workers per queen is between 2 and 3, a similar group size to the most common one in nature. (Adapted from Bartz and Hölldobler, 1982.)

relative productivity of the co-founding queens and eliminate all but the most productive, they would be maximizing the chance that they preserve their own mother. The most productive queen will have laid the most eggs and have most daughter workers, so that most workers will be favouring their own mother and their own inclusive fitness.

Perhaps the reason that pleometrosis very rarely gives rise to stable long-lasting polygyny is that the queens who come together to start a nest are unrelated, or very distantly related, so that the resulting colony is an unstable mixture of competing family factions. This leads to the death of the extra queens or one or more divisions of the colony. For example, in the common meadow ant, *Lasius flavus* in England, queens also come together into groups to start colonies. In this species the survival of queens in groups was much higher than the survival of individual queens, but these pleometrotically-formed colonies often split into groups with each queen going off with her own workforce once the first stages of colony growth had been passed (Waloff, 1957).

1.4.1 Secondary polygyny

Permanent polygyny almost always occurs through the adoption of fertile females into an established colony. This secondary and permanent polygyny is even harder to explain than primary polygyny. As more queens are added to the nest there will be a decrease both in the average relatedness of the members of the colony, and in the inclusive fitness they gain by raising the sexual progeny of the extra queens.

A queen that allows another to enter the colony and share its resources is being an altruist, and for such altruism to spread by kin selection the ratio of the gain (B for benefit) to the beneficiary over the loss (C for cost) to the altruist, in terms of offspring produced, has to be greater than the reciprocal of the relatedness (r) between the altruist and the beneficiary, i.e. $B/C > 1/r$. The more closely related the queens the less the B/C (benefit/cost) ratio has to be, and the greater is the chance of polygyny developing and being maintained. We should therefore expect multiple queens to be close relatives.

Craig and Crozier (1979) estimated coefficients of relatedness in polygynous nests of the Australian jumper ant, *Myrmecia pilosula*. The average relatedness of queens was estimated with 95% confidence limits to be 0.243 ± 0.113; the average relatedness of workers was 0.172 ± 0.053. In both cases these relatedness coefficients are significantly different from zero, showing that the queens are relatives. However, it is clear that the queens were not all full sisters ($r = 0.75$). The low average relatedness among the queens within colonies suggests that colonies adopt extra queens that are a mixture of more or less related individuals such as cousins and grandaughters as well as daughters of the original queen or queens.

Pamilo and Varvio-Aho (1979) found that from year to year the average coefficients of relatedness between workers in one particular nest of *Formica sanguinea* varied between 0.42 ± 0.098 and 0.311 ± 0.124. These values were similar, as one might expect, to the coefficient of relatedness between the female brood produced by the colony and its workers. In addition to female brood, males were also collected: their relatedness to the workers was unexpectedly high at 0.6016 ± 0.402. The study colonies were nesting in very stony terrain, so de-alate established colony queens could not be recovered. Nevertheless these data support the idea that *F. sanguinea* colonies typically have several related, singly-mated queens per nest with one queen possibly dominating the egg-laying. The very high relatedness between the male brood and the workers has to be interpreted with caution, however, because the calculated figure had a very high standard error. These data may be accounted for by workers sometimes either laying many of the male-producing eggs themselves or raising nephews, i.e. sons of sister queens.

Pearson (1982, 1983) examined intracolony relatedness of workers and queens in polygynous colonies of the common British ant *Myrmica rubra*. The coefficients of relatedness that he estimated for both workers and queens were extremely low, and in fact were not significantly different from zero except in certain populations in certain years. Furthermore he re-examined the data of Craig and Crozier (1979) using more advanced statistical techniques and concluded that the coefficients of relatedness among *Myrmecia pilosula* queens were also not significantly different from zero.

We can conclude that the relatedness of queens in polygynous nests is apparently often quite low and therefore polygyny is hard to explain unless there is a very great gain (B) to the adopted queens in terms of their reproductive potential and a relatively small loss (C) to the established queen or queens. If, as Craig and Crozier (1979) have suggested, colony founding is especially difficult, established queens might increase their genetic output by accepting daughters (or other relatives) back into their colony, thus decreasing their personal output by increasing their offspring's (or other relatives') output. For example, if the relatedness of queens in

Myrmica is taken as 0.243, the benefit to the adopted queen must be about four times the loss to each of the established queens that permit the adoption.

Some species in the *Rhytidoponera impressa* group of ants have both monogynous and polygynous colonies in the same geographical area. The monogynous colonies are headed by a queen who is larger than the workers and had a winged stage in her development. However, in polygynous nests the reproductive females resemble workers morphologically, but have the ability to mate and store sperm from matings, so that they can produce both inseminated diploid and uninseminated haploid eggs. In *Rhytidoponera confusa* and *R. chalybaea*, Ward (1983) showed that within neither species was there evidence of reproductive isolation between the two types of colony. In the monogynous colonies, genetic markers suggested that the single queen mates only once and is the mother of all colony members, including males. The relatedness of workers is roughly 0.70 and the relatedness of workers to males about 0.2. In the polygynous nests the mated 'workers' are often closely related and share reproductive output fairly equally, so that in spite of polygyny the relatedness among the workers in these nests is about 0.3 and the relatedness of workers and males is roughly 0.16. Uninseminated laying workers were found in about a quarter of the polygynous nests. In ecological terms there appears to be a balance of advantages between the two types of nest; in wetter and more productive habitats the monogynous colonies appear to be more efficient and productive, but the worker-reproductive colonies are better at dominating the habitat, by colony budding, in particularly dry environments or in less stressful but extremely patchy habitats.

The fitness of colony members in monogynous and polygynous colonies of the *Rhytidoponera impressa* group is also affected by the sex ratios these nests produce. Recall that monogynous and polygynous nests often occur in the same habitat, and that there is no reproductive isolation between these forms. Where the two types of colony occur together, monogynous nests produced extremely female-biased sexual broods: the proportional investment in females was 82%. In polygynous colonies more males were produced and the investment in female reproductives was relatively much less, sometimes as low as 36%. This investment in females is extraordinarily low even though it includes, as part of the resources given to new 'queens', the workers that accompany new reproductive females when they form a secondary colony that buds off from the parental one.

The biased investment in favour of females by monogynous nests was positively correlated with both the relative frequency and the density of sympatric polygynous nests. How monogynous colonies are able to assess the relative abundance of polygynous colonies, and hence compensate by producing more females to maintain the best sex-allocation strategy, is as yet unknown. In certain habitats polygynous nests are completely absent. In one such case the estimated proportional investment in female reproduction by the monogynous nests was 0.76. This suggests that, in the absence of external factors, workers in monogynous nests control the sexual investment ratio. Thus in monogynous *Rhytidoponera* colonies workers may maximize their inclusive fitness by controlling the investment patterns of their colonies, while in polygynous colonies uninseminated workers may also enhance their inclusive fitness by producing their own sons in quite large numbers: uninseminated laying workers were found in about a quarter of polygynous colonies.

A conflict over sexual investment apparently also occurs in the tiny acorn ant, *Leptothorax longispinosus*, in North America. These minute ants often nest in already existing cavities in plant material, such as acorns that have been hollowed out by weevil larvae. Perhaps, in part as a consequence of the availability of these nest sites, colonies of this species are not only highly polygynous but also highly polydomous, literally occupying many homes. This makes an interesting situation because workers in satellite-nests in the absence of queens may be more free to fulfill their own preferences in terms of sexual investment to a greater extent than workers who are in the physical presence of queens in queen-right nests. Herbers (1984) was able to show significant differences in sexual investment between nests with queens and the satellite-nests without queens; the latter's broods were much more female-biased, as might be expected if kin selection was acting on the workers to promote their own inclusive fitness to the detriment of the queens'.

One particularly interesting situation for comparing monogynous and polygynous forms of the same species occurs in populations of the imported fire ant, *Solenopsis invicta*, a species which has caused economically important damage over a wide area of the deep south of the United States, following its rapid spread over the last 80 years. Until recently true functional polygyny was unknown in this species, but subpopulations with this trait have arisen apparently independently and on numerous occasions within the last 15 years. In the singly-mated monogynous form, worker nestmates are revealed by electrophoretic techniques to have coefficients of relatedness statistically indistinguishable from 0.75 (Ross and Fletcher, 1985). *Solenopsis* workers do not possess ovaries, and their obligate sterility means that all the males are produced by the queens. In the polygynous colonies the queens are also singly inseminated, but remarkably nestmate queens are no more closely related to one another than they are to any other queens in the population. Ecological differences between the two forms are hard to pinpoint, as polygynous forms may occur either together with or separately from the monogynous form. One apparent difference is that the workers of the polygynous form are much smaller than the monogynous form. In this case it is hard to believe that kin selection has had a major role in the evolution of polygyny in this species, and it is possible that this is one of the apparently rare occurrences in which mutualism among colony-founding queens has given rise to permanent polygyny.

Polygynous colonies can reproduce by budding rather than always being started by workerless queens. Workers that help to form new colonies may increase their inclusive fitness by helping to ensure the survival and reproduction of the queens they produce. The importance of colony reproduction by budding in polygynous populations is demonstrated in studies in which the genetic relatedness of individuals from neighbouring nests has been compared. For example, if colonies reproduce by budding there should be strong tendency for neighbouring nests to be more closely related than more distant ones. Such a trend has been shown in another species of *Rhytidoponera* in which only polygynous worker-reproductive nests occur. In *Rhytidoponera mayri* relatedness was low but significant both within and between neighbouring nests, as is consistent both with multiple gynes and with reproduction by budding (Crozier *et al.*, 1984). Since in *Rhytidoponera mayri* neighbouring nests are related (albeit fairly distantly) one might expect to find some reduction in the normal territorial hostility between neighbours (see Chapters 5 and 8) and perhaps some positive cooperation. This may occur only to the extent of a reduction in fighting between foragers from neighbouring nests. Such a reduction in hostility between related neighbouring nests is also seen in wood ants. This is seen in an extreme form in a 'super-colony' of the ant *Formica yessensis* on the Ishikari Coast of Hokkaido which has been estimated to contain 306 million workers and 1 080 000 queens living in 45 000 interconnected nests which dominate a territory of 2.7 km^2 (Higashi and Yamauchi, 1979) (see Chapter 5).

In this phenomenon of habitat domination through the budding of polygynous colonies, coupled with reduced hostility between neighbouring nests, we can detect the first stages of one of the most interesting manifestations of polygyny in ants. This is the evolution of a unicolonial form of social organization in which there are apparently no colonial boundaries and nests can grow indefinitely through the re-adoption of queens, and queens rarely if ever form colonies alone. This form of colony organization, in species such as Pharaoh's ant *Monomorium pharaonis*, *Paratrechina longicornis*, and *Iridomyrmex humilis*, has enabled them to dominate certain habitats: indeed, they have become major pests both in natural communities and in public buildings. Since their colonies possess many queens and their nests are highly diffuse, they are virtually indestructible. In such species, ant colonial structure has become an extended family in which powerful colony-level selection, in the context of unusual ecological conditions, seems to have overridden the selfish interests of nest members.

1.4.2 *Polygyny as a form of parasitism*

A number of authors have suggested that polygyny in many species of ants may be a form of parasitism. For example, Elmes (1972) has suggested that the large number of queens found in *Myrmica rubra* colonies are essentially parasites on the workforce. More queens are present than are needed to produce all the eggs that the workers are able to raise, and such queens represent an energetic drain on the colony. One can therefore consider the queens and workers of these colonies as almost independent populations with the queen population growing partly at the expense of the future growth of the worker population. For this reason the growth of both populations may be limited in a logistic manner by their own growth and that of the other caste. Elmes therefore used a modified host–parasite population model to make predictions about the structure of these colonies. The model suggested that the various numbers of queens in the different colonies should fit a log-normal Poisson distribution, as indeed proves to be the case. Such a distribution of these 'parasitic' queens might arise if newly inseminated queens return to the ground after the nuptial flight and enter colonies partly haphazardly and partly as a function of the existing size of the colonies.

Intriguingly, in many *Myrmica* species there are two distinct queen morphs; a large form known as a macrogyne and a small form, the microgyne. Recently, Pearson (1981) (see also Pearson and Child, 1980) has shown by electrophoretic techniques that these different queen morphs in *Myrmica rubra* are almost certainly separate species with the small form parasitic upon the larger. In the microgyne queens found in various *Myrmica* host species there is a progressive trend towards increasing parasitic qualities, including a reduction in size and a reduced tendency to produce workers. Probably polygyny has been important in this way in the evolution of the parasitic workerless ant queens known as inquilines (Buschinger, 1970; see also Chapter 7).

1.4.3 *The behaviour of polygynous queens*

Although, as we have discussed above, secondarily adopted queens may be a separate species from their host, in most cases the multiple queens within a colony are all the same species. Even here, if they are distantly related we might expect to find behavioural evidence of conflicts between them.

Wilson (1974) studied the behaviour of the multiple queens in colonies of another North American species of tiny acorn ant, *Leptothorax curvispinosus*. He suggested that there was strong competition among these queens. There are two sources of food for queens; the first is from trophallaxis with workers and the second from the larvae who probably possess digestive enzymes not found in the adults and therefore can function as part of the social stomach of the colony. When workers were not feeding queens they tended to retreat before them, clearing a path to the larvae where the queens compete among themselves for larval secretions. *L. curvispinosus* queens do not appear to recognize one another as individuals, nor do they form dominance hierarchies as occur among certain ant workers (see section 1.5). The queens recognize their own eggs, however, and roughly handle and eat the eggs of the other queens with the result that there are marked differences in the reproductive rate of the various queens within the nest. Evesham (1984) conducted a somewhat similar study on colonies of *Myrmica rubra* with multiple queens. In these colonies there are marked differences in the behaviour of the different queens; some are largely stationary and are rarely aggressive to other queens, whilst subordinate queens seem to be highly unsettled and move restlessly throughout the colony often initiating attacks on one another. The stationary and subordinate queens appear to be ranked in a dominance hierarchy which future studies may show to be related to their reproductive potential.

In *Myrmica rubra* colonies most if not all the males are the product of eggs that have been laid by workers (Smeeton, 1981). Perhaps this is an adaptation on the part of workers to salvage some inclusive fitness by substituting their own sons for those of queens that are at best very distantly related to the workers. Very little is known about the behavioural interactions between egg-laying *Myrmica* workers. It is possible that there is fierce competition among such workers for the role of being the major egg layers.

1.4.4 Oligogyny

Though there may be reproductive competition among queens in polygynic colonies this is rarely associated with direct aggression between them; instead they tend to remain mutually tolerant and tend to cluster together (Hölldobler and Wilson, 1977). However, some colonies with multiple queens are best described as being oligogynous. As in a human oligarchy, rule is by a few; also as in the human situation, there is often overt rivalry among the rulers. Oligogynous colonies are characterized by worker tolerance to more than one queen and by antagonism among the queens. The result is that the queens tend to be widely dispersed through the nest. Hölldobler and Carlin (1985) studied the behaviour of oligogynous queens in rapidly growing colonies of the Australian meat ant, *Iridomyrmex purpureus*. This species seems to have almost every conceivable form of colony foundation: most colonies are founded by individual queens (haplometrosis), but others are formed by budding from large colonies or by a few queens starting colony life together. In the latter case, when queens are forced to live together due to the small size of their colony, they have ritualized bouts of antennal boxing. By this means one queen asserts her dominance and this is associated with greater egg production by the highest-ranking queen. Once the colony has grown to a size at which the queens can live separately, the queens disperse to different parts of the nest where they attain equal status and produce approximately equal numbers of eggs. The adaptive significance of dominance behaviour among the queens is not yet clear, as in the case studied they achieved equal status before any sexuals were produced. However, in other circumstances it is possible that the dominant queen could inhibit the reproduction of her subordinates almost permanently, or take over the majority of the colony when it buds to form a new one.

1.5 Dominance hierarchies in workers

Two recent studies have shown that there are competitive interactions among workers in monogynous colonies over egg-laying rights. Where there are multiple, distantly-related queens there is likely to be particularly strong selection on workers to retain the ability to produce eggs that will become males. Worker egg-laying, even in monogynous species, is much more common in ants than the popular view of sterile workers suggests. Recently, Cole (1981) showed that workers of *Leptothorax allardycei* produce 22% of all the eggs laid in a colony with a queen and so it is likely that they will be contributing a large fraction of all the males produced by their colonies.

By individually marking workers of *L. allardycei* and observing the tiny colonies of this species under a dissection microscope, Cole (1981) was able to deduce the status of workers from their ritualized antagonistic interactions. Certain workers were often seen to dominate others by lunging at their subordinates and later pummelling the body of the submissive worker with their mandibles. Typically, submissive workers withdraw their legs and antennae and crouch perfectly still while they are being bullied in this manner. Sometimes dominant workers actually climb on top of their lower-ranking nestmates. On other occasions the subordinates simply run away when a high-ranking worker moves towards them. Such interactions are not only highly ritualized but also highly predictable. If for example x dominated y and y dominated z, then x was also sure to dominate z; there were very few reversals in these linear hierarchical rankings.

High-ranking workers receive more food than the low-ranking workers in the nest, have greater ovarian development and probably produce more sons than their subordinate sisters. Dominance hierarchies in *L. allardycei* occur in both queen-right and queenless colonies, but antagonistic interactions occur among workers at a greater rate in the absence of the queen probably because, without the influence of the queen, workers are free to produce many more eggs. The queen's influence over her workers is probably pheromonal, as Cole (1981) did not observe any direct behavioural role of the queen in these hierarchies.

The tiny slave-maker *Harpagoxenus americanus* (which enslaves both *Leptothorax longispinosus* and *L. curvispinosus*) also has linear dominance hierarchies (Franks and Scovell, 1983) (Table 1.2). In this species the queen is actively involved in behaviourally dominating her workers. Dominance interactions between *H. americanus* females are particularly highly ritualized, as might be expected among ants that have potent morphological adaptations to demolish other ants during slave-raids. Rather than engage in fighting that might injure other colony members and reduce the inclusive fitness of dominant workers by lowering the productivity of the colony, the slave-makers ritually drum their antennae over the bodies of subordinates, later actually standing on and over them, as in *L. allardycei*.

Slaves are the only source of food for the slave-maker workers as they do not forage themselves. High-ranking slave-makers are fed more often by the slaves, have greater ovarian activity and probably produce more sons than their subordinates. The proximate reason for the dominance behaviour is clearly seen when the high-ranking workers interrupt trophallaxis between

SOCIAL BEHAVIOUR AS A SELFISH STRATEGY

Table 1.2 Dominance hierarchy tabulated from 10 h of observation of a *Harpagoxenus americanus* slave-maker colony (from Franks and Scovell, 1983). Each entry is the number of interactions between the indicated pair of ants.

Dominant ants	Subordinate ants						Total times dominating	Ovarian status	
	Queen	P	GA	T	GR	B	W		Eggs	Active ovarioles
Queen	—	7	9	1	3	3	4	27	—	—
P		—	41	54	28	32	49	204	5	5
GA		1	—	22	24	26	24	97	4	4
T				—	18	27	38	83	2	3
GR				2	—	22	16	40	1	3
B						—	7	7	0	0
W						1	—	1	1	1
Total times dominating	0	8	50	79	73	111	138			

Table 1.3 Dominance hierarchies in a colony of *Harpagoxenus americanus* before the removal, in the absence, and after the return, of the dominant worker 'M' (from Franks and Scovell, 1983).

Dominant ants	Subordinate ants					Times fed by slaves
Initial 27 hours' observation						
	Queen	M	L	F	O	
Queen	—	41	25	13	7	90
M		—	75	40	21	98
L			—	29	18	50
F				—	2	55
O						54
Queen fed by slave-maker workers		8	0	0	1	
After removal of M, 10 h observation						
	Queen		L	F	O	
Queen	—		20	8	5	28
L			—	24	24	18
F				—	10	13
O					—	21
Queen fed by slave-maker workers			1	1	1	
After return of M, 10 h observation						
	Queen	M	L	F	O	
Queen	—	8	3	3	0	31
M		—	48	39	9	49
L			—	29	4	15
F			3	—	10	14
O				3	—	9

lower-ranking slave-makers and slaves. In effect they highjack extra food so that they have enough energy to produce more of their own eggs. The queen breaks into these interactions and maintains her position at the head of the hierarchy. Most remarkably, the only trophallaxis between slave-makers is when the queen is fed by one of her workers, predominantly the highest ranking slave-maker worker. This could be an adaptation on the part of the queen to reduce the energy reserves of the worker that most threatens her own fitness.

This hypothesis is supported by an experiment in which the dominant worker was removed (Table 1.3). In the absence of the alpha worker, interactions between the remaining workers increased significantly in frequency and were particularly violent between the beta and gamma workers; also, the queen solicited food for the first time from the beta and gamma workers. When the original alpha worker was replaced she immediately approached the beta worker that had usurped her place and had an extremely prolonged and violent battle with this worker. The original alpha worker regained her dominant position, and once again the queen demanded and received food predominantly from this highest-ranking worker. One other intriguing aspect of these dominance hierarchies is that it is the lowest-ranking workers that take the greatest risks with their lives, as these workers do most scouting for slave-raids.

In this way conflicts over the production of sexuals in ant colonies can have marked effects on the tasks that workers are prepared to undertake (see Chapter 4) and how altruistic they are in serving the rest of the colony.

1.6 Cooperation and competition

This chapter has emphasized the extraordinary puzzles presented by the evolution of sterile or nearly sterile workers in social insects. We have shown how the enigma of sterility is associated with the greater reproductive success of relatives and can therefore be partly explained by the theory of kin selection. Hymenoptera, including ants, are highly susceptible to kin selection because they have a haplodiploid form of sex determination which leads to asymmetric and very high degrees of relatedness among certain members of their family units. Haplodiploidy and the adaptations present in almost all female hymenopterans for parental care are both important factors that have favoured the evolution of workers and a colonial way of life. Social life is not all cooperation, however; conflicts too are involved in the evolution of ant societies through kin selection. Such conflicts are likely to occur not only between the queen and her workforce over the sex ratio produced by their colony, but also over who produces the males. Conflicts of interest are even more prevalent in ant colonies that have multiple queens or multiply-mated single queens. All these conflicts place constraints on the overall 'factory' efficiency of ant colonies.

There are sometimes conflicts between workers and larvae and among larvae. In most ants it is the workers who determine whether female larvae will develop into workers or queens (see Chapter 3). In *Myrmica* colonies part of the control of queen production involves workers biting and nipping larvae apparently to inhibit their development into queens (Brian, 1979). In this situation there may be a real conflict of interests between workers and larvae.

In the more primitive ants, such as the ponerine *Ambylopone* or *Myrmecia* and *Nothomyrmecia*, the larvae are much more autonomous than in more advanced species. Such larvae can crawl around their own nest and at least to some extent control how much food they can obtain. Given all the other conflicts in ant societies, it would almost come as a surprise if such larvae did not themselves compete for food and even perhaps have dominance hierarchies.

CHAPTER TWO

THE PHYLOGENY OF ANTS

Eusociality is found today in two insect orders, the Isoptera or termites, all of which are eusocial, and the Hymenoptera, which has eusocial forms among the wasps, the bees and the ants. Termites, like ants, are essentially insects of the soil, and like them have wingless workers. The Isoptera show many important differences from social Hymenoptera: they do not have haplodiploid sex-determination and perhaps as a consequence their sterile castes include both males and females, unlike the female-based societies of the Hymenoptera. Unlike ant larvae their immature forms have legs and relatively hard cuticle, and in some groups are able to act as workers or pseudo-workers. We shall meet termites again as an important food resource for tropical ants.

2.1 The origin of ants

2.1.1 *The ant family tree*

Ants share the haplodiploid method of sex-determination with all Hymenoptera. Hymenoptera have also a marked tendency to parental care. This means that resources are used to promote the survival and growth of existing young rather than to produce a larger number of offspring. Among non-social groups, time and energy may be spent in placing eggs in the tissues of plant hosts (sawflies) or host insects (parasitic forms), or in the production and provisioning of a nest (non-social bees and wasps). The existence of this tendency to parental care has undoubtedly been important in the evolution of social Hymenoptera. It is not easy, however, to trace the development of ant sociality from pre-existing parental behaviour patterns. Non-social bees and wasps have many varieties of parental care, and eusociality seems to have evolved at least eight times independently among

Table 2.1 The sub-families of ants (based on Taylor, 1978).

Sphecomyrminae
Poneroid complex
  Pseudomyrmicinae
  Myrmeciinae
  Ponerinae
  Dorylinae
  Ecitoninae
  Leptanillinae
  Myrmicinae
Formicoid complex
  Nothomyrmeciinae
  Aneuretinae
  Dolichoderinae
  Formicinae

them. Ants, by contrast, form a single superfamily, the Formicoidea, with limited, structural variety—conventionally all ants are included in a single family, the Formicidae, within the superfamily Formicoidea—all of whose members are eusocial. The living members of this family are usually placed in a dozen or so subfamilies, which are listed in Table 2.1. Because we have to classify some 12 000 species inside this single family, smaller taxonomic groupings are often used, especially the division of subfamilies into tribes such as the Ecitonini and the Dacetini. The structural features all ants share are: (i) the ability to produce both winged and wingless females; (ii) a specialization of the first segments of the abdomen to form one or two nodes; (iii) an antenna with a long first joint (the scape) and a larger number of smaller joints (the funiculus); (iv) mandibles meeting medially in an extended medial margin, with a number of subapical teeth (Figure 2.1). There are also special features of the wing venation in sexual forms, and all ants possess a pair of metapleural (more properly propodaeal) glands just anterior to the waist. These characters force us to look for the ancestral pre-ant among the smaller, rather primitive, families of the Aculeate (i.e. those Hymenoptera with a sting rather than an ovipositor), and not in the larger groups of solitary wasps such as the Sphecoidea or Vespoidea. In any case the parental care given by ants shows some important differences from the methods of the solitary and social wasps. Ants do not isolate their young by laying each egg in a separate cell, stocked with prey, as the social wasps and many solitary forms do. Instead, ant's eggs are laid in a heap or egg-pile inside one or more quite spacious brood chambers. Malyshev (1968) argued from this that the ancestral pre-ant laid a large number of eggs on a large

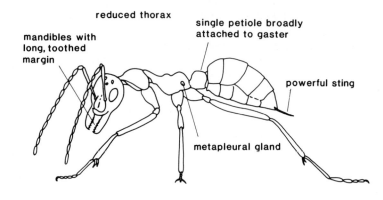

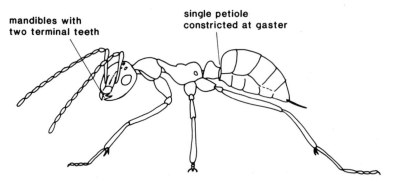

Figure 2.1 The main features of a schematic primitive worker ant. (Above) a primitive ant as predicted on general grounds before the discovery of *Sphecomyrma*; (below) *Sphecomyrma* itself. The two discrepancies are starred. (Adapted from Wilson *et al.*, 1967.)

prey (perhaps a beetle larva inside rotting wood as *Scleroderma* (family Bethyloidea) does), and then tended them as a group. Anatomical arguments, on the other hand, would place early ants nearer to *Methocha* (Tiphiidae). Most of the behavioural features of these lower Aculeate groups, like Bethylidae and Tiphiidae, are very specialized, however, and throw little light on the origin of ant behaviour.

2.1.2 A Mesozoic fossil ant

Fossil ants are not uncommon in amber, most of which is Tertiary in age, and yet they belong to genera which are still in existence today. The exception is the Cretaceous *Sphecomyrma freyi*, found in New Jersey in 1966

THE PHYLOGENY OF ANTS

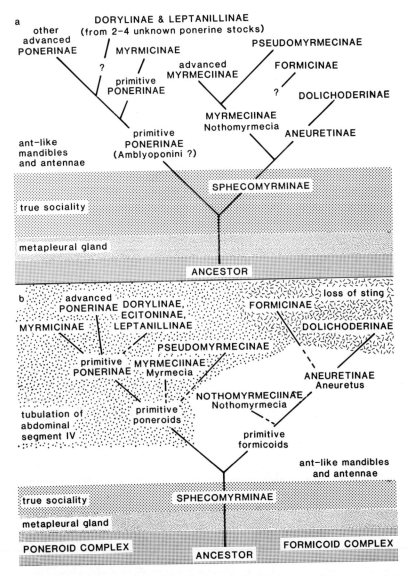

Figure 2.2 The family tree of ants. (Above) a hypothetical cladogram taking account of the discovery of *Sphecomyrma* (after Wilson *et al.*, 1967); (below) a modified cladogram taking account of the special features of *Nothomyrmecia* (after Taylor, 1978.)

(Wilson et al., 1967). The two known specimens of this species are the oldest Aculeate Hymenoptera dating back 100 million years, and since they are clearly wingless workers, they are the earliest known social insects. Yet *Sphecomyrma* is in many ways quite close to living Tiphiidae like *Methocha*. *S. freyi* has a single well-formed node, somewhat ant-like antennae and propodaeal glands. Rather unexpectedly, its mandibles are pointed and wasp-like. This combination of characters caused something of an upset, since many 'living fossil' ants, especially the Australian *Myrmecia* and *Amblyopone*, have ant-like mandibles and although they have less well-defined nodes there is a distinct tendency towards a second node (Figure 2.1). Possibly *Sphecomyrma* and *Amblyopone* should therefore be placed at the base of two different branches of the Formicoidea: the myrmecoid complex, based on *Sphecomyrma* and containing the bull-ants like *Myrmecia* and the more widespread Formicinae and Dolichoderinae, and the poneroid complex based on *Amblyopone* and some other Ponerinae, including the Myrmicinae and the Dorylinae (Figure 2.2a).

2.1.3 *A living fossil ant*

As with many other groups of land animals, the relatively early isolation of the Australasian land mass allowed the survival of some interesting living fossil ants. The large and aggressive bull-ants *Myrmecia* quickly impressed themselves on nineteenth-century explorers of the interior. They live in large earthen nests as predators in semi-arid areas. Two workers of a much less common and quite distinct type, *Nothomyrmecia macrops*, were found in 1934, but in spite of several expeditions no further specimens were found until 1977 (Taylor, 1978). *Nothomyrmecia*, like *Sphecomyrma*, forces us to change our views of the evolution of the ants. *Nothomyrmecia* is not as large or as robust as the bull-ants, and has a single rather slender node; that is, there is no constriction between the third and fourth abdominal segment, so that it superficially resembles a more delicate Formicine, like *Oecophylla*. The most significant difference, however, is in the structure of the fourth segment (Figure 2.3). In *Myrmecia*, *Amblyopone* and all the more advanced members of the Myrmeciinae, the tergite and sternite, which form the dorsal and ventral sides of the segment have become 'tubulate', that is their most anterior parts are joined and are withdrawn inside the preceding segment. This allows a controlled rotation and telescopic extension, and probably facilitates the use of the sting; it may have led to the development of a fully formed second node (postpetiole) as in Myrmicinae. In *Nothomyrmecia*, however, the tergite and sternite of segment 4 are freely articulated, like the

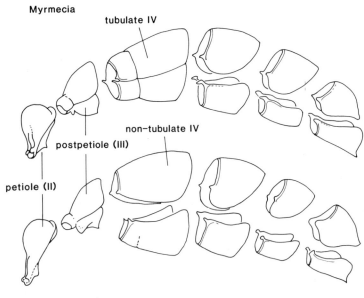

Figure 2.3 Diagrams of the abdominal plates of two primitive ants, showing the differences in segment IV, which has its tergite and sternite separate in *Nothomyrmecia*, but joined to form a tubulate structure in *Myrmecia*. (After Taylor, 1978.)

succeeding segments, as they are in the Formicinae and Dolichoderinae. This led Taylor (1978) to put forward a different scheme for the relationships of the main subfamilies of ants (Figure 2.2b). *Sphecomyrma* can be placed at the base of the tree, since the significance attached to single or double nodes has altered. Above this the tree divides into two branches. The ponerine complex, with *Amblyopone* and other primitive Ponerinae near its base, includes all the forms like the Myrmicinae and Pseudomyrmecinae which have two nodes and a tubulated fourth segment, as well as the Dorylinae in which the workers at least have double nodes of a sort. The formicoid complex, with *Nothomyrmecia* near its base, contains the single-noded Formicinae and Dolichoderinae, and all its members have the fourth abdominal segment untubulate.

2.1.4 *Adaptation to liquid feeding*

The importance of the non-tubulation of the base of the abdomen lies in its freedom to expand as the crop within is filled. The Formicinae and the

Dolichoderinae, the most advanced families of the formicoid complex, have both undergone marked adaptation to feeding on liquids. The best-known feature of this adaptation is the development, apparently independently in the two subfamilies, of methods of retaining liquids in the crop (Eisner, 1957). Many Hymenoptera possess a crop, formed as an in-line dilatation of the posterior region of the foregut. As in most insects, the junction of this region with the midgut forms the proventriculus. This controls the passage of material from the crop to the midgut. The contents of the crop can be emptied through the mouth by regurgitation, and fed to other members of the colony. This food-traffic, or trophallaxis, is of great importance in the transmission of food from foraging ants to the queens and to nurse ants who transmit it onwards to larvae (see Chapter 3). Food which passes through the proventriculus to the midgut is presumably assimilated, though in fact little is known of the physiology of the digestive system in ants. In more primitive ants and in all the poneroid complex, the proventriculus has a cross-shaped lumen, lightly sclerotized, which can be closed by the active contraction of muscles (Figure 2.4). In the subfamilies Formicinae and Dolichoderinae, this active retention of crop contents is replaced by a passive dam. The anterior surface of the proventriculus, inside the crop, is developed into lobes (sepals or quadrants) that are held shut by the pressure of liquid in the crop. How the liquids are pumped into the crop is not clear: the only known mechanism would be peristalsis of the oesophagus. In formicine ants at least the proventricular dam is so effective that external pressure on the gaster normally results in the discharge of the crop contents through the mouth, and this can be used to assay crop content in the field. The expansion of the crop is facilitated by the free movement of the tergite and sternite of segment 4. The non-tubulation of the abdomen in *Nothomyrmecia* suggests that early in evolution the ants split into two groups, one with a tubulated abdomen and an effective sting, the other untubulated and able to carry large amounts of liquid in its crop and tending to lose the sting. The climax of this latter trend is the development of replete ('honey-pot') workers in both Dolichoderinae and Formicinae; these have enormously enlarged gasters, and function as colony storage reservoirs for liquids. This type of trophic lifestyle is apparently impossible for ants of the poneroid complex which, as we shall see, have either retained a dependence on solid insect flesh, or developed alternative lifestyles based on the collection of seeds or the culture of fungi. The queens of Dorylinae have expanded abdomens, due not to the crop but to the growth of the ovaries. This physogastry does involve the separation of the sclerites of the fourth segment, and these queens, unlike their workers, have only a single node.

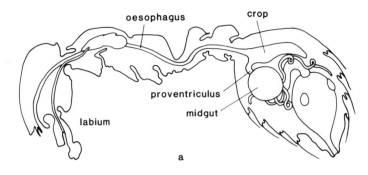

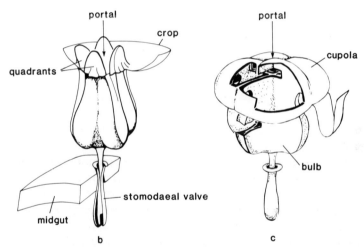

Figure 2.4 The modifications of the gut in ants for feeding on liquids: (*a*) the essential parts of the alimentary canal in *Myrmica*; (*b*) the proventricular valve in *Myrmica*; (*c*) the passive dam in *Iridomyrmex detectus* (Dolichoderinae).

2.2 The subfamilies of ants

2.2.1 Myrmeciinae

This subfamily is confined to Australia and New Guinea. They vary in size from large slender 'inch-ants', up to 25 mm long, to the smaller 'jumpers', down to about 5 mm long and mostly diurnal. Many are brightly coloured, presumably as a warning of their formidable sting.

2.2.2 Ponerinae

These ants have a single node but many have a constriction of sorts between segments 3 and 4. The group is in fact quite diverse, although most species are more or less committed to a predatory life. Robust forms, like *Rhytidoponera* and *Odontomachus* search on the surface for prey or carrion. Very large forms such as *Paltothyreus*, *Leptogenys* or *Megaponera* make group raids on termite nests. Less robust genera are cryptic predators in soil or humus (*Amblyopone* and *Ponera* itself), while some specialize in raids on the nests of other ants (*Cerapachys*). Some species, however, have more varied habits like the African *Brachyponera* which is a predator in the wet season but feeds on seeds in the dry season.

2.2.3 Dorylinae

These are tropical group raiders known as driver ants in Africa and army ants in the New World. They have recently been reviewed by Gotwald (1982). The most closely studied genus *Eciton* has very large colonies (0.5 million), and the workers have a very large size range. Each colony has a single queen, which is wingless and undergoes physogastry (enlargement of the gaster) to accommodate very large and productive ovaries. Egg production is periodic and related to a two-phase behaviour cycle. In the nomadic phase, during which the brood are all larvae, there is no permanent nest. The colony migrates each evening to a new temporary bivouac, an above-ground cluster of workers, which contains the brood and the queen. The colony spends the night in the bivouac. The pupation of the brood precedes a new pulse of egg-laying by the queen, and the colony changes to a statary phase. In this phase the colony remains in the same site until the eggs and the pupae hatch. In both phases the colony makes raids into the surrounding forest, driving insects out of cover, capturing them, and carrying them back to the brood and queen. In some *Eciton* this periodicity is very regular, for instance the nomadic phase of *E. hamatum* lasts 15–17 days and the statary phase 18–23 (Schneirla, 1971). In other species like *Anomma*, although there is some regularity in raiding and emigration, these alternating nomadic and statary phases are not found (Gotwald, 1982). *Eciton* and the Northern *Neivamyrmex* raid more on the surface than most other Dorylinae. Gotwald (1982) points out that Dorylinae should not simply be divided into epigaeic (surface hunting) and hypogaeic (subterranean) species. A species may be hypogaeic both in its nesting and in its foraging (like many species of *Dorylus* and some *Aenictus*) and this is probably the primitive condition. Others, like *Anomma*, forage above ground from a hypogaeic nest. Yet others, like the *Eciton* species which were studied earliest, both nest and forage above ground. The males are also rather curious, and were not at first recognized as belonging to the same species. Their biology is not well known, but they seem to seek out raiding columns and return with them to the nest to mate with the less mobile queens. It is a principle in ecology that primary consumers generally have larger populations than secondary and higher level consumers, so that predation would not be expected to be a good basis for supporting large colonies. The Dorylinae are the only predatory ants which breach this rule: other ants which maintain large colonies do so by a mixed alimentary strategy, in which other non-predatory sources of food, such as honeydew, seeds or fungi, supplement the predatory diet, or in some cases replace it.

2.2.4 *Pseudomyrmecinae*

These are a small and specialized group which live in hollow stems, or in specially developed parts of plants. The Central American *Pseudomyrmex* live in the large thorns of *Acacia* trees, which they hollow out. Once a colony has become large its workers patrol the whole of the tree, killing or chasing away other insects. Like all the subfamily they have formidable stings and can even repel browsing deer. In addition they bite the growing tips of any other plant which touches their own tree. In the secondary scrub where these *Acaciae* grow this frees the tree from competition and is a considerable advantage. The ants and their tree thus live in a state of mutualism, from which both benefit (Janzen, 1966). The plants in return provide specially developed, highly nutritive organs which the ants collect.

2.2.5 *Myrmicinae*

This very large subfamily live in a wide variety of ways, and unlike the families mentioned so far are common in temperate as well as tropical habitats. Because of the tubulation of the fourth abdominal segment, and the limits on the crop which this has imposed, the Myrmicinae have not, with a few exceptions, diversified their resource base by exploiting liquid foods like honeydew or nectar. This is not to say that myrmicines do not collect honeydew, for many species do so, but rather that it forms on the whole a less important part of their diet than in Dolichoderinae and Formicinae. Familiar European ants like *Myrmica* and *Tetramorium* feed on insect prey as well as collecting honeydew and some seeds. They play a very large part in the trophic structure of meadows (Petal, 1978) and heaths (Brian, 1983). The small ants in the tribe Dacetini have a much greater specialization; their jaws and behaviour are modified for the capture of small soil insects, like *Collembola*. In the more specialized forms like *Daceton*, the mandibles are elongated; while hunting the ant abducts them, so that they lie at right angles to its body. Sensory hairs project from the ant's labrum; when these touch prey the mandibles are released and close very rapidly on the prey, before it can jump away. A similar 'trap-jaw' mechanism is also found in *Odontomachus* (Ponerinae). Other dacetines like *Smithistruma* have shorter jaws and depend on stealth to catch prey.

Although many generalized ants, like *Tetramorium* and *Myrmica*, feed on seeds, they are probably not well able to utilize the starch in seed reserves as an energy source. Some seeds, such as *Viola* and *Cyclamen*, have a special organ, the elaiosome, on the outside of the seed, which is rich in oils. Ants collect the seed and remove the oily part, abandoning the remainder of the seed, which is thus dispersed. This is called myrmecochory (see Chapter 6), and is to be distinguished from the use of a much wider range of seeds as a major food resource by harvesting ants. It is a habit typical of desert, semi-desert and Mediterranean habitats. The best-known harvesting ants are myrmicines, *Messor*, and *Monomorium* in the Eastern Hemisphere and *Pogonomyrmex*, *Veromessor* and *Aphaenogaster* in American deserts. Their workers have large jaws and powerful muscles, and in *Messor* and *Pheidole* are dimorphic. The latter is a worldwide genus, living on seeds as well as insects. All its members have a marked dimorphism of workers, with soldiers or majors, in which the head is disproportionately large compared to the rest of the body, and smaller workers or minors, of normal proportions. The two 'subcastes' undertake different tasks in the colony (see Chapter 4). Traditionally, seed collection has been seen as the accumulation of winter stores, and an example of the wisdom of ants compared with grasshoppers. The father of ant ethology, Pierre Huber, poured scorn on the fable that harvesting ants even existed, believing that this referred to the collection of nest material by ants like *Formica rufa*. Huber was Swiss, and less well-informed about the habits of Mediterranean ants than the writers of scripture and fable. Mackay and Mackay (1984) recently pointed out, however, that the stores of *Pogonomyrmex* do not seem to be used during the winter in New Mexican deserts, since the amount stored is not lowest at the end of the winter. They believe that the stores are used throughout the year whenever active foraging has to cease, for instance because of heavy predation.

Myrmicine ants are not all limited to nesting in the soil or in humus. Many slender species, like *Leptothorax*, live in hollow twings under bark or inside galls or in nuts such as acorns on woody plants. Other slender species live inside the nests of other ants (*Myrmicoxenus*), or as thief ants which penetrate the galleries of other species and steal food or brood (*Solenopsis fugax*). *Solenopsis invicta* on the other hand is an aggressive fire ant, which collects seeds and tends aphids as well as insects, and uses a most painful sting to defend its nest sites against trampling farm stock and farmers. More robust forms like *Cataulacus* and *Zacryptocerus* live inside wood, apparently in deserted beetle tunnels, etc. Other arboreal ants, particularly in the genus *Crematogaster*, are an exception to the rule that Myrmicinae, with their more primitive crops, are not major collectors of honeydew, for many *Crematogaster* species collect it avidly. Two interesting facts go with this exception. First, the function of an enlarged crop is to transport liquids over a distance, so it is interesting that the most prominent myrmicine exploiter of honeydew has moved its nest to the trunk or twigs of the tree which is the ultimate source of the honeydew. Second, *Crematogaster* is characterized by a curious articulation between the second node and the fourth abdominal (= first gastral) segment. This is usually said to allow the ant to lift its gaster over its body dorsally and apply chemical defence substances to enemies. The sting is not well developed, and special glands in the legs provide trail pheromones which in other Myrmicinae are laid from the glands of the sting. The effect of these anatomical peculiarities on the function of the crop might be worth investigating.

The other group of myrmicines which has apparently escaped from the limits of a predatory existence forms the tribe Attini, the fungus or leafcutter ants (Weber, 1982). The description refers to the habit of some species which cut leaves from living plants and carry them to the nest: some other species collect dead leaves or caterpillar faeces. In either case the material forms a substrate (cellulose plus nitrogen) for specific genera of fungi, which are 'cultivated' on a fungus garden within the nest. The garden is also the site where the queen and brood are kept, and provides their food. Attine ants are exclusively American in distribution, and probably evolved from tropical seed-collecting forms, as South America drifted away from Africa during the Tertiary period. Later the connection with North America allowed them to spread as far north as New Jersey (40° N), as well as southward as far as Argentina (44° S), and to altitudes of up to 3000 m in South America. The group probably evolved in tropical forest, surviving in isolated patches of forest during the warm and arid periods of the

Pleistocene. At this time perhaps the grass-feeding species of *Acromyrmex* and *Trachymyrmex* arose, as well as the desert-scrub inhabitants like *Trachymyrmex smithi*. Leaf-cutter ants hold a strange ecological position: since they collect plant material directly they must be thought of as herbivores. Their dependence on fungi to process this plant material is not unique, and can be compared with the situation of the higher termites or of the ruminants. The leafcutting species *Atta and Acromyrmex* are remarkable for the wide range of plants from which they remove leaves. They take young leaves of almost any tree in their area, usually concentrating on one or a few genera at any time, but switching to others as that source becomes exhausted or too mature. This has led to the suggestion that attines are able to evade the chemical defences of plants by relying on the ability of their fungi to adapt to changes in the substrate and to produce a uniform feedstock for the colony. The identity of the fungi used is hard to establish, since identification depends on production of fruiting bodies, and these are never formed during culture by ants. It has been hard to grow the fungi in culture, and cultures are susceptible to infection with extraneous fungi. It seems probable that the fungi are close to, or members of, *Lepiota*. The fungus is eaten by the ant and about 50% of its dry weight is available as a soluble nutrient to the ants as carbohydrates 27%, amino acids 4.7%, protein-bound amino acids 13% and lipids (including sterols) 0.2%. The ant provides the fungus with its substrate and with a stable nest climate, for exposed mycelium dries rapidly and dies if exposed to the air. The ants also apply their faeces to the fungus gardens. This recycles nitrogen as allantoin and allantoic acid, free ammonia and amino acids. The contents of the rectal sac also have high levels of protease activity. Since in artificial cultures the fungus grows poorly when supplied with nitrogen in the form of proteins, but much better when these proteins are first hydrolysed, it might seem that the ants supply the protease which makes nitrogen from the leaf substrate available, as amino acid, to the fungus, thus adjusting the carbon/nitrogen ratio. It appears, however, that these enzymes are produced not by the ants but by the fungus. The role of the ants is in moving enzymes from mature parts of the fungus garden to sites of inoculation, so that the fungus can begin to utilize its substrate more rapidly. The ants also free the fungus from competitors, possibly by the action of a number of aromatic acids liberated from the metapleural glands. The result of this biochemical cooperation is the rapid cycling of plant material (Weber, 1982). Although leaf-cutting may seem to the peasant farmer to be a damaging activity, it almost certainly has a generally beneficial effect on the environment.

2.2.6 Dolichoderinae

One of the two 'higher' subfamilies of the formicoid complex (Figure 2.3b), these are linked to the primitive *Nothomyrmecia* by a relict form *Aneuretus simoni*. Like *Nothomyrmecia* itself, *Aneuretus* has a very restricted distribution, and is known only from humid forest in Sri Lanka; again like *Nothomyrmecia* it was lost at its original site, but found again in some numbers in 1955. While *Aneuretus* resembles the Dolichoderinae in its thin, collapsible cuticle and in its wing venation, it has an elongated, low node like *Nothomyrmecia*, and a powerful sting. It also lacks any of the rancid odours possessed by many Dolichoderinae. It has therefore a mixture of primitive and advanced features, which earn it its own subfamily, the Aneuretinae.

The true Dolichoderinae are a rather compact group anatomically. They have no sting, and rely instead on a battery of chemicals produced by glands in the anal region. The opening of these glands is usually just ventral to the tip of the gaster, and is a transverse slit, not circular as in the Formicinae. The glands often have a large number of products, much studied in the cosmopolitan *Iridomyrmex humilis*. This use of chemical weapons seems to be the reason for the success of some very small Dolichoderinae, not only *I. humilis*, which has been spread by human activity to many parts of the world, but for instance *I. pruinosus* and *Conomyrma insana* of North American deserts. These ants, only 3–4 mm long, swarm from their multiple nest entrances to tend scale-insects, membracids and aphids on desert shrubs. Although only some of the shrubs carry these Homoptera, nearly every bush that does have them is dominated, at least in daytime, by one or other ant species. No other diurnal ant is able to profit from them. This domination of trees is taken further in the more tropical *Azteca*, whose relations with trees resemble those we described for Pseudomyrmecinae. The Central American *Cecropia* is a tree of regenerating secondary bush, which develops hollow stems, inhabited by *Azteca*, and also provides a food source from Müller's organs below each leaf stalk. *Azteca* is apparently able to colonize many trees if they have hollow green nodes or stems where the ants can tend Homoptera; a large colony can even dislodge the much larger *Pseudomyrmex triplaridis* from its obligate host tree *Triplaris americana*. Not all Dolichoderinae are small, however. The Australian meat-ant *Iridomyrmex purpureus* workers are up to 10 mm long, diurnal, and dominant over other ants on the soil surface. Other larger species like *Liometopum* are dominant on plants in rocky desert hills in Mexico, where they tend scale insects, etc., on xerophytic plants. Their behaviour and habits recall those of the less aggressive species of *Formica* or *Lasius* in cooler habitats.

2.2.7 *Formicinae*

This last subfamily (the Camponotinae of some authors) are a fairly homogeneous group. Like the Dolichoderinae they have lost the sting; it is replaced by a device which in many species can squirt a mixture of substances, often including formic and other acids in high concentration, to a distance of several centimetres to repel enemies, or to kill prey. The opening of this structure is at the tip of the gaster, where there is usually a circular pore, sometimes called the acidopore, often surrounded by a circlet of bristles. The subfamily is well adapted to feeding on honeydew, and all forms have an efficient passive dam formed by sepals projecting forward into the crop from the proventriculus. In the northern parts of both the Old and the New World the genera *Lasius*, *Acanthomyops* and *Formica* are important. The first two (*Acanthomyops* is now used only for some of the American forms) live by tending Homoptera. Some, like the European *Lesius flavus* and all *Acanthomyops* are subterranean; *L. flavus* builds mound nests in soil and can be locally dominant on grazed grassland. Others like *Lasius fuliginosus* are adapted to life in trees where they build carton nests (see Chapter 3). *Formica* has several divisions, often used as subgenera. Species like *F. fusca* make medium-sized colonies; since they are 'enslaved' by other *Formicae* they are placed in the subgenus *Serviformica*. In America species of the subgenus *Neoformica* like *F. pallidefulva* are similarly enslaved, as well as species related to *F. fusca* like *F. neoclara*. The 'slave-making' species (*F. sanguinnea* and its American relations) have some anatomical peculiarities and are placed in *Raptiformica*. *Polyergus* is an example of a more specialized slave-maker (see Chapter 7). Most of the members of *Formica sensu strictu* are wood ants. They build large mounds of vegetable litter, which are responsible for a nest temperature some 5° above soil temperature. By a combination of aggressive expansion of the colony into several polydomous nests, and a diet of honeydew and insects collected mostly on trees, they dominate many areas of broad-leaved and coniferous forest in Northern Europe and Asia. American forms seem slightly less aggressive. In the Palaearctic *Formica* shows a marked zonation from south to north: *F. fusca* is replaced by *F. lemani* in Northern Europe and on mountains, and *F. lugubris* and *F. aquilonia* similarly replace *F. rufa* and *F. pratensis*. *Formica* evidently does not spread readily to the arid or semi-arid lands of the Mediterranean or the deserts of the

American Southwest, where it is replaced by *Cataglyphis* and *Myrmecocystus* respectively. *Myrmecocystus* is known as the 'honey-pot ant' because of the development of some of its workers into repletes. In these liquids are accumulated in the crop until the gaster is grossly enlarged, to four or five times its initial linear dimensions. It is usually assumed that the liquid is honeydew, but it appears that it can also be prey juice or even water (Snelling, 1981). A similar habit is known in Australian *Melophorus* and to some extent in the Mediterranean *Proformica*.

The largest of ant genera is *Camponotus*, with over 1000 species. The genus has been divided into a number of subgenera, mainly on anatomical grounds. Most *Camponotus* can be recognized by the low saddle-like thorax and the origin of the antennae well behind the clypeus. Carpenter ants (*C. herculeanus* and *C. pennsylvanicus*) are large ants able to bore in the softer summerwood of living trees. They are found throughout the Holarctic except in the British Isles. Some other species, like the subgenus *Colobopsis*, live in tunnels in wood and have a major worker with a round head which is used to block the tunnel entrances. Other species live in a wide variety of ways in wood, soil, etc. They can range in size from small to very large, and their workers very often show a very marked size range with polymorphism. *Oecophylla* is the genus of the well-known weaver ant, ranging with some variation from West Africa to North Queensland. The weaver ant is arboreal, and builds a very diffuse nest through one or more trees, using silk produced by third-stage larvae to 'sew' living leaves together. The African and Asian *Polyrachis*, with two pairs of spines on the thorax and a pair on the node, also uses silk in this way; some species are arboreal, but others live among stones in arid places where the silk is used to seal the nest.

2.2.8 *Further reading*

This survey of ant diversity has necessarily been very limited in scope. Further details of particular faunas and their natural history can be found in Brian (1977) (British Isles), Michener and Michener (1951) (American social insects), Creighton (1950) (North America, but becoming dated), Greenslade (1979) (South Australia).

CHAPTER THREE

ANT ECONOMICS

Ant colonies are in many respects closely analogous to human factories; a division of labour is organized so as to maximize profits, portions of which are reinvested to maintain the production machinery, sustain further growth or set up new factories. An ant colony producing more sexual offspring than its neighbours is effectively increasing its market share, just as a profitable company might do.

An entire branch of microeconomics is devoted to the theory of the 'firm', defined as any entity using economic inputs such as land, labour, and capital to produce outputs such as goods and services. The 'economizing problem' facing the firm is that of deciding how much output to produce, and how to use the various inputs, given the technological relation between outputs and inputs and the price of inputs and the value of output. The translation of this economic definition to ants is a simple one. Inputs such as land and labour are equivalent to the territory and the workers respectively, whilst the capital represents stored resources such as the seed banks of harvester ants, or the wing-muscle reserves and fat-body of colony-founding queens or overwintering workers. The profitable output is purely the colony's sexual offspring. The value of the ouput is calculated in terms of inclusive fitness. As explained in Chapter 1, the queen and workers might put different values on the male and queen offspring of the colony. Even this can be seen as analogous to conflicts between management and labour in human factories. In the ant situation such conflicts must be taken into account when predictions are made of the economic strategy of colony members over their lifetime.

One of the most striking analogies between human factories and ant colonies concerns economies of scale. Ant colonies of intermediate size, just like firms of medium capacity, often grow faster and are more efficient *per capita* than both smaller colonies (or companies that are especially prone to bankruptcy), and very large colonies (or firms), where inefficiencies arise from stretched and broken lines of communication which cause slow responses to changing environments.

3.1 Economies of scale

For simplicity ant colony growth can be considered to pass through three distinct stages (Oster and Wilson, 1978).

3.1.1 *The colony-founding stage*

Colonies may be founded by a single queen (haplometrosis) or groups of queens (pleometrosis). In either case young colonies tend to end up with only one queen: long-lasting polygyny is generally associated with adoption of queens by established nests (see Chapters 1 and 7). The reason queens may come together at the colony-founding stage is to pool their resources during the first and most vulnerable stages in colony growth. This is like partners pooling their capital to start a company. The capital available to a queen is in the form of the energy from her fat-body or wing muscles, redundant once the mating flight is over. This is generally the only energy available to queens as, with the exception of some ants such as primitive ponerines, most ant foundresses avoid the risks of foraging for food and wall themselves into their nest. This is called claustral colony foundation.

The queen or queens use their energy reserves to raise their first brood of workers. Workers of this first generation are noticeably smaller than those from mature colonies and are called nanitics, literally dwarf workers. Not only are nanitics tiny but they are also characteristically timid, and quickly run away from danger where workers from mature colonies would wade into battle. This timidity is clearly adaptive, as the death of nanitic workers would result in the death of the queen and colony through starvation.

Colony-founding *Atta* queens produce exactly the appropriate size of new workers that are just large enough to be able to accomplish all the necessary tasks of leaf-cutting, tending the fungus garden and raising brood but are as small as possible so that as many as possible can be produced with the limited resources available to the small colony (Wilson, 1983*b*; see also Chapter 4).

3.1.2 *The ergonomic stage*

After the first slow and tentative steps in colony growth, once the nanitics have raised a second and larger brood of bigger workers, colonies typically

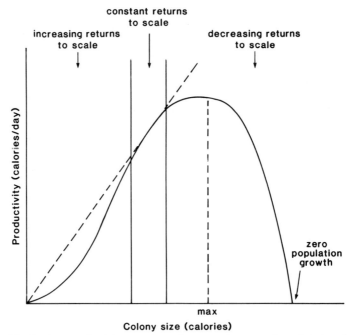

Figure 3.1 A colony productivity curve that relates net colony growth as a function of colony size. As the colony population increases, the total net yield of growth (measured in calories) can be expected to change in a way that conforms to a general pattern but varies among species. This hypothetical curve suggests that most social insect colonies will pass through growth phases involving increasing returns, constant return to scale and then decreasing returns to scale. The points of maximum colony growth and zero population growth fall within the period of decreasing returns to scale. (Redrawn from Oster and Wilson, 1978.)

enter a period of extremely rapid growth. This has been called the ergonomic stage (Oster and Wilson, 1978) in which the colony engine works most productively. Typically colonies grow like many other logistic populations (see Figure 3.1): the ergonomic stage approaches the inflection point of the logistic curve where the worker population grows most rapidly so that the colony growth trajectory to this point shows so-called increasing returns to scale. In this stage, therefore, profits are re-invested in workers of the most efficient type and in infrastructure such as the nest. For example *Atta* colonies in the ergonomic phase produce a wider range of worker sizes to construct more efficient production lines with more specialized and sophisticated castes performing more and more precisely designated tasks (Wilson, 1983*a*; Chapter 4). In the ergonomic stage colonies are transformed from cottage industries to production-line assembly plants.

3.1.3 *The reproductive stage*

At some critical size (near or just beyond the inflection point) a colony must begin to produce sexual offspring in order to realize its inclusive fitness. This involves shunting some of the energy made available by the increasing workforce into the rearing of male and female larvae. The timing and quantity of such production is critical to the future fitness of all colony members. Earlier reproduction has its advantages as successful daughter queens will sooner be able to produce their own daughters (who are of course granddaughters of the original queen and contribute to her colony's inclusive fitness). However, the advantages of early reproduction can be offset by the reduction in the total number of sexual offspring a colony can produce in its lifetime. If a colony puts energy into sexual production when it is small the growth rate of the worker population will be correspondingly reduced, and less energy will be available for sexual reproduction in the future. These trade-offs between early but limited reproduction or delayed but more productive reproduction make assessing the optimal investment schedule for a colony very complex. This problem in economics is increased by the seasonal changes in resources. Most ants rear sexual larvae over one or more seasons and then release them at a single time of year. For this reason they face all the classic economic problems of seasonal industries. Ice-cream factories must maintain some staff and machinery all the year round, but one bad summer can spell disaster.

3.2 Colony life-history strategies

3.2.1 *The schedule of growth, investment and reproduction*

As the definition of the 'firm' emphasized, one of the factors involved in determining economic planning horizons is the technological link between inputs and outputs. In ant terms this involves such factors as (i) the longevity of workers, and queens; (ii) how much food workers bring in; (iii) how efficiently they rear offspring; (iv) how long worker and sexual larvae take to develop; (v) the mechanism of control of female caste determination; (vi) who controls this switching to gyne production; and (vii) who produces the males. Out of the 12 000 or so living ant species, the number for which even the majority of these parameters are known probably does not even enter double figures. So most models of colony life-history strategies are purely generalized and often have limited predictive value for real cases. One of the few genera in which most of the above parameters have been measured is *Myrmica*. This is largely due to the dedicated work of M. V.

Brian and his students and colleagues. Brian et al. (1981) have constructed predictive models that incorporate realistic parameters for *Myrmica* colonies. These models can be used to explore possible adaptations in the growth and reproductive schedule of colonies, and to predict the best, i.e. the optimum, compromise between producing some sexual offspring earlier or more later. The model incorporated the following optimization criteria.

(i) The worker population must, after maintaining itself, produce as much food as possible.
(ii) The worker population structure and the schedule of food utilization should produce an investment ratio of males and females that maximizes inclusive fitness. (In *Myrmica* workers produce all the males so that according to Trivers and Hare (1976) colonies should invest equally in males and queens (see Chapter 1).)
(iii) The fecundity of queens and workers should be sufficient to use all the available food at the colony size determined by conditions (i) and (ii), so that neither food nor eggs are wasted.

These constraints were applied to a model that describes the flow of resources through a *Myrmica* colony as indicated in the box diagram in Figure 3.2. For example, the amounts of food that workers bring in and use themselves or feed to larvae were represented by equations whose numerical parameters were estimated by observations and measurements on colonies in the field and laboratory. In this way a numerical model colony was set up, and run to investigate at what size colonies should switch to producing sexual offspring to maximize their inclusive fitness.

According to the model the colony grows logistically and then passes into a period of stability or convergent oscillations (for example, large numbers of sexuals are produced by colonies only in alternate years). Oscillations could be caused by time lags in the system, or by competition between sexual brood and worker production, but also to some extent by the way density-dependent processes were approximated in the model equations. Such oscillations are, however, known in some real ant populations. Damping, which results in smaller and smaller oscillations, was also a feature of the model, caused by the limited egg production and by the way in which food was distributed in proportion to the brood requirements. For example, the model mimicked the likely real situation in which queens are the first to be fed by workers, then the larvae are given all they require, after which the queens take all the remaining food. This realistic arrangement tends to ensure that the food is partitioned appropriately between egg production by the queen and larval growth.

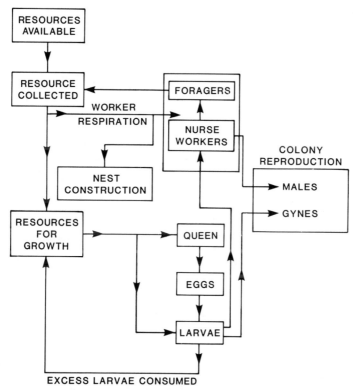

Figure 3.2 Schematic flow sheet for a colony of *Myrmica*, considered as a 'firm' producing gynes and males. The amount of resource available depends on the territorial area and its richness; the amount collected increases with the number of foragers, but is subject to diminishing returns. Some of the resource collected is respired by workers as they forage or build infrastructure such as the nest. The remainder is available for colony growth; the queen has the first option on this, and she uses it to produce eggs. What she will not take goes to larvae, and is used for growth. Larvae in excess of what can be supported starve and their substance is recycled. When the ratio of workers to queen becomes large, the 'firm' produces its final products: males from haploid eggs laid by workers and gynes from diploid eggs laid by the queen. These leave the colony to mate, and they, and workers that die, are lost to the system. (Based on Brian *et al.*, 1981: a detailed description and numerical model will be found in this paper.)

The model had some extremely interesting characteristics. First the model yielded the highest inclusive fitnesses for colony members when the parameter representing food input was set very near that recorded in a natural grassland habitat. This suggests that the switches which determine schedules of sexual production in real colonies have been set by natural selection to favour the greatest fitness of colonies. In other words, colonies are highly adapted to their particular natural habitats.

In the models the switches corresponded to the known behaviour of *Myrmica* queens. Generally workers have a tendency to bias female larvae to become gynes rather than more workers. The queen opposes this: first, by laying worker-biased eggs when young; second, by forcing workers to overfeed young larvae and so initiate early metamorphosis which results in worker development; and third, by limiting the feeding of older larvae. Though the workers produce all the males the queen also has some control over this production. In the intimate presence of the queen, workers tend to produce trophic rather than viable eggs. In this way queens are able to recover food that was not released by the workers and use it for their own offspring. Queens also control male production by varying their own oviposition rate and by supervising the egg piles in the nest.

In addition the model predicted that colonies in poor habitats, i.e. those where food input is extremely limiting, should specialize in producing males because these sexual offspring are much smaller and less expensive than queens. This prediction also fits observations of colonies in poor habitats, many of which produce only males. Indeed, this is true not only for *Myrmica* but also for ants of many other genera.

3.2.2 *Knowing when to split*

The vast majority of ants produce winged sexuals that mate away from the parental nest and start colonies without the initial help of workers. However, in certain species, reproduction occurs through the fission of the parental colony. In this way the new queens are given part of the workforce of the old colony. Colony fission should be seen as distinctly different from colony budding—the latter occurs almost as a process of vegetative growth when parts of an already polygynous colony form new satellite-nests that retain their links to the original nest. Colony fission occurs when a strictly monogynous colony momentarily re-adopts a newly inseminated daughter queen who almost immediately leaves with part of the workforce to form an entirely distinct new colony.

Colony fission is only likely to evolve at all in circumstances in which solitary queens would have an almost vanishingly small chance of survival. Fission involves producing a very low number of daughter gynes, at massive cost, because each takes a large number of workers, instead of producing thousands of winged offspring.

Colony fission occurs in all army ants including the phylogenetically distinct Old World (*Dorylus*) driver ants and *Eciton* in the New World. In terms of the economic policy of army ant colonies we should pose four questions. (i) At what size should a parental colony split? (ii) What portion of its workforce should it give to daughter colonies? (iii) How many daughter colonies should it produce at any one splitting event? (iv) At what time of year should it undergo fission?

These questions are based on the assumption that natural selection will have favoured colonies with the best policy so as to maximize their Darwinian fitness, in terms of reproduction and survival, relative to other army ant colonies in the same population. Seen in these terms colonies should split as quickly and as many times as possible, subject to the constraint of not jeopardizing their own survival or that of their daughters. Survival probabilities, colony growth and splitting rates should all closely correlate with colony size in those species which show reproduction by fission. This means that colony size is the common currency in which the best economic policy can be calculated.

Simple mathematical models show that a colony should always divide into just two parts and at such a size that the combined growth rate of the daughter colonies just exceeds the growth rate of the parental colony at the point of division. This rule can be understood intuitively. A colony will be favoured if it produces daughter colonies as quickly as possible but at such a size that the daughter colonies can also go on to produce daughter colonies as quickly as possible. By following the formula 'split evenly at the point where your daughters will grow fastest compared with your own growth' a colony will have the best chance of maximizing the number of granddaughter and great-granddaughter colonies it produces. In other words, it should maximize its market share. Not surprisingly, other animals that reproduce by fission, from flat-worms to sea anemones, follow these very same economic life-history rules (Franks, 1982c; 1985).

The problem of when to split in relation to seasonality can also be understood intuitively as well as mathematically, although the result may at first seem surprising. In order to obey the above splitting rule a colony may have to grow to a size at which its efficiency is dropping off, i.e. the part of its growth trajectory where it suffers decreasing returns to scale. It may at that point be able to split into smaller colonies that are altogether more efficient, i.e. have increasing returns to scale. In this case small is not only beautiful but also more resilient. If seasons of shortage and plenty are predictable a colony may be favoured if it divides into the smaller, more efficient, units to face the bad times. An alternative approach comes to the same conclusion. If there are highly productive seasons alternating with less productive ones, why waste the best times in reorganizing rather than maximizing profits? An ice-cream firm would hardly choose to re-organize in the summer when it could do so with less loss of profits in the winter. Similarly, certain sea anemones divide during the winter when food for growth is in shortest supply.

These predictions for the growth and reproductive schedules of colonies that reproduce by fission have been tested in the case of the army ant, *Eciton burchelli*, on Barro Colorado Island, Panama. *E. burchelli* colonies are particularly good subjects for this kind of study because their growth occurs in particularly regular and measurable stages. These colonies maintain 35-day brood production schedules. They produce new broods in distinct batches so that all the larvae grow during the same 15-day period, pupate in synchrony and eclose as new workers 20 days later. During the process of pupal development a colony remains in the same statary bivouac site. When the new workers eclose, the army ants drop the discarded pupal cases on the ground and begin an emigration to a new bivouac site. This means that the amount of brood they have raised over the previous 35 days can be estimated by collecting the pupal cases, and the colony size can be estimated by filming the emigration. In this way, Franks (1985) was able to determine colony growth rate as a function of colony size. Then it was possible to estimate how fast two daughter colonies would grow supposing that colonies of various sizes split equally in two. From these graphical relationships between parental and combined daughter colony growth rates, the optimum size at division could be estimated. The predicted optimum corresponded closely to the size at which colonies were seen to split (Figure 3.3). Such colony reproduction takes place only in the dry season in Panama which, as predicted, is the season of slowest growth when the insect prey of army ants are least abundant.

Colony reproduction in army ants has a fascinating natural history. Over the 35 days leading up to the division, a colony of *Eciton burchelli* produces a brood of approximately 3000 males and only 6 queens. In the last statary day the colony produces two raid systems in opposite directions and then the new queens and the original single parental queen run down these raid systems. When a queen reaches the end of a raid system a new bivouac is formed. Only the two most vigorous queens succeed, one going in each direction. The extra queens, which may include the original colony mother, are held back by workers at the old bivouac site, where they are abandoned and left to die. Eventually the workforce of the original colony splits equally with half accompanying each of the successful queens. In addition half of the new worker larvae and half of the males that were produced by the

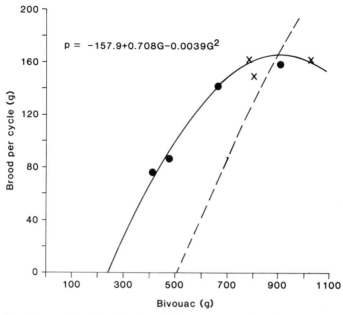

Figure 3.3 Grams of brood produced per 35 days (P) as a function of bivouac dry weight (G grams) for colonies of the army ant *Eciton burchelli*. Circles indicate colonies that produced worker broods, and crosses indicate colonies that produced sexual broods (see Table 3.1). The solid line shows a polynomial, least squares, curve-fit based on all 7 colonies ($r = 0.99$, $N = 7$, $p < 0.001$). The broken line was derived from the formula of the solid line and represents the combined growth rate of two identical daughter colonies as a function of the size of their parental bivouac. The intersection of the solid and broken lines indicates the optimum size for colony division. (Redrawn from Franks, 1985.)

original queen are taken to each new colony. The new colonies go their separate ways and during the first 15-day nomadic phase the males fly off to attempt to find other colonies of their species which they must enter to inseminate queens. All army ant males are very large and closely resemble their conspecific queens in having very long cylindrical gasters. In all of the separate phylogenetic lineages of army ants, males have decidous wings, a very rare trait in ants in general. Males also superficially resemble queens in that they have batteries of exocrine glands on the surface of their gasters that produce similarly attractive secretions in similar sites to those produced by queens. Both males and females are surrounded by attentive worker entourages as they move through the colony. Recently, Franks and Hölldobler (1987) have suggested that the superficial resemblance of male to conspecific queen in all army ants is due to the fact that workers are involved in selecting both the mothers and fathers of new army ant colonies and that in this extraordinary process of vicarious sexual selection both males and females have fastened on the same channels of communication to demonstrate their vigour and desirability to the workers.

3.3 The flow of resources within the colony

The strategies of resource allocation described above depend on pathways by which resources pass from foragers to other members of the colony, and eventually to the final product—a new generation of sexual forms. In Chapter 5 we shall describe the systems of discriminators and visas which restrict this flow of resources to colony members. First, however, we look at the actual patterns and mechanisms of resource flow among the members of the colony. Although solid pieces of prey may simply be dumped inside the nest or near larvae, liquid food is usually passed from the crop of one ant to the mouth of another by regurgitation. This behaviour is very obvious in laboratory colonies, particularly when they have just been given liquid food. Wheeler (1910) elevated the behaviour, under the label trophallaxis, to the most important feature of social life: today perhaps we see it as a means by which sociality, and in particular the propagation of kin, is achieved.

3.3.1 Food exchange between workers

The expression 'food exchange' is inaccurate, but probably too firmly established as a description of the movement of material through the colony to be corrected now. In fact in any 'exchange' it is usually very clear that one ant is the donor and the other the recipient. The donor probably has a visibly enlarged gaster; she extrudes a droplet of liquid from her mouth on to her mandibles, which are held open; she keeps her antennae more or less still and widely spread. The recipient on the other hand keeps her antennae in constant movement, and may also beat on the head of the donor with her front legs. In *Myrmica* the recipient may stridulate (Lenoir, 1972; Wallis, 1962). The exchange can be initiated in several ways. In colonies which have been starved and then fed abundantly, workers with full crops may regurgitate spontaneously, either to a recipient or even on to the floor of the nest. In this situation the recipient may get food without any previous antennal contact. More usually the exchange takes place after a long or short episode of antennal 'solicitation' in which one ant, usually the donor, taps or strokes the frons of the other with her antennae. These movements have been described in detail based on high-speed film by Lenoir (1973). The continual antennal movements of the recipient seem to have the main function of orientating the recipient correctly as in the honeybee (Free, 1956), and the donor may regurgitate without any palpation by the would-be recipient. Usually, however, the recipient has to maintain a certain level of excitement to initiate regurgitation and subsequently to maintain it. Equally, the donor may use one or even both antennae to maintain the exchange. As Lenoir says, 'The continuation of the trophallaxis needs therefore an equilibrium between the reciprocal stimulations of the donor and of the recipient'. The exchange ends when this equilibrium is broken on the initiative of either of the participants. In some cases the donor will try to break away from the recipient, perhaps because her crop is becoming empty. Sometimes the recipient prolongs the bout by changing her position or intensifying her palpation. On the other hand the recipient may use her front legs to push herself away from the donor, who also can sometimes in spite of this maintain the exchange by palpating the recipient with her antennae. Comparison of the signals involved with those in wasps (Montagner, 1963) and bees (Montagner and Pain, 1973) suggests that three sorts of signals are involved; a demand signal from the recipient to the donor, a signal of acceptance from the donor to the recipient, and a signal of termination. The signals are apparently much less specific in ants than in wasps, and Lenoir concluded from an informational study of the signals, and the ants' responses to them, that in ants the signals of one participant are not good predictors of the behaviour of the other; behaviour is primarily determined by the 'motivation' of the respondent. Bonavita-Cougourdan (1984) agreed with this and concluded that the flow of liquid from the donor was not in any way dependent on the antennal movements of the recipient.

The stimulatory equilibrium probably models an underlying physiological equilibrium, based perhaps on the degree of crop distension of the two participants. The flow of food from donor to recipient continues until the difference between them is reduced. In some cases the direction of flow appears to reverse at the end of the exchange (Bonavita-Cougourdan *et al.*, 1979), and this too is perhaps evidence of equilibration. We know too little of the physiology of feeding in ants to expand this explanation at present. In *Dolichoderus quadripunctaus* antennal contacts can lead to the emission of a droplet from the cloaca of the donor which is licked up by the recipient (Torossian, 1973); the significance of this is not known.

3.3.2 *The flow of food to larvae*

In some predatory species such as large ponerines or *Myrmecia*, solid food (whole prey or pieces of prey) are brought into the nest and dumped. In

Myrmecia the larvae are mobile enough to move to this dump under their own power, and have extrusible heads with which to bite into the prey. In *Aphaenogaster subterranea* workers place solid food on the heads of larvae, or may carry the larvae to the food pile. Larvae are taken one at a time from the larval pile, placed on the prey for about 30 min and then removed. In experiments a larva could only feed if it was placed with its head in contact with the food. If its trunk was in contact, or it was 5 mm away, it did not feed in a 2 h test. *Myrmica rubra* workers chew food before feeding it to larvae (Brian, 1973), flies are sucked dry and the residue deposited in the nest. Many ants separate the solid and liquid parts of prey by pressing the solids into the infrabuccal pocket within the mouth; the liquids are passed down to the crop. Other components, particularly lipids, can also be separated into the pocket and seem to be transferred from there to the postpharyngial glands. *Monomorium pharaonis* deposits dried grains of food from the infrabuccal pocket in its nest, but is not known to feed the grains to larvae.

In many species, however, crop contents are regurgitated to larvae. The crop contents may be honeydew, but expressed prey juices are also fed to larvae in this way. Larvae make small movements of their mouthparts, but it is not clear if these have any effect in encouraging workers to feed them. The regurgitated food probably contains glandular secretions as well. In *Tapinoma erraticum* it is particularly the young workers 10–15 days old that regurgitate to larvae, and they regurgitate much more rarely to other workers (Lenoir, 1979). This may be associated with the development of postpharyngeal and other glands in the youngest workers. Naarman (1963) and Markin (1970) found that radioactively marked glandular secretions were transferred to larvae from workers, and Buschinger and Kloft (1973) found that a group of young *Monomorium pharaonis* workers surrounded the queen, and acted as 'poison-tasters', shielding her from poison baits by passing food through their glands. The passage of food to the glands took several days. Another source of food for larvae is the trophic eggs laid by workers; in *Myrmica rubra* young workers with functional ovaries lay eggs which have thin shells and are apparently not viable. These are used to transfer material from the body tissues of workers to larvae. This route is particularly important also in the incipient nest. The queen of *Tetramorium caespitum* lays about 200 eggs before the appearance of her first daughter workers, but there are never more than 20 larvae in the colony; only about one larva in 50 develops into a worker in this phase of colony development (Poldi, 1963).

3.3.3 *The course of food flow in the society*

If the pathways of food exchange depend on an equilibrium of crop distension without dominance, as has just been suggested, it would provide the basis of a mechanism for the diffusion of resources evenly to all parts of the colony. This would not be what the colony needs or what the models outlined in section 3.2 require, however. If the colony is a factory producing new sexuals, resources must flow preferentially to the queen or queens (who produce eggs) and to growing larvae, with preference among them to those which will become sexuals. The last preference needs delicate control, however, since production of sexuals must be delayed, in accordance with the schedule of investment. In terms of the very simple model which is emerging this means that queens and larvae must act as resource sinks, that is to say they must receive far more resource than they pass on. (The same is true of adult males in those ants like *Camponotus herculeanus* and *C. ligniperdus* in which adult males overwinter in the nest (Hölldobler, 1966).) They could act as sinks either by having no crop mechanism for giving food back as in larvae, or by moving food rapidly from the crop and assimilating it in the midgut as must happen in the phenomenally productive queens. However, both queens (Lenoir, 1973) and males (Hölldobler, 1966) regurgitate to workers when their own crops are full.

The actual course of food can be traced on a fairly slow timebase by means of radioactively marked food (Naarman, 1963; Markin, 1970; Buschinger and Kloft, 1973), or more rapidly by behavioural observation. Lenoir (1979) applied the methods of multivariate statistics to a behavioural study of this kind. He recorded the frequency with which individually marked workers in a laboratory colony of *Lasius niger* were seen in the act of giving or receiving food with other workers, feeding larvae of various sizes, feeding the queen, carrying larvae or prey, etc.; a total of over 30 different behaviours. By using a computational technique called correspondence analysis Lenoir was able to detect how each behavioural act was associated with every other act (e.g. how often each was, or was not, carried out by the same worker at different times), and so to separate the worker force into groups which specialized, at least temporarily (see Chapter 4), in different groups of acts. There were great differences between individual workers; in particular some were more often donors and other recipients in exchanges with other workers, and others fed mostly larvae. In this way three classes of workers could be distinguished. Foragers had highly dilatable crops, well-developed acid glands and atrophied Dufour's glands and ovarioles. They readily left the nest to forage and to retrieve larvae left outside the nest. They never received food from nestmates and never nursed larvae. Nurses in contrast never emerged from the nest, remaining permanently with the brood which they fed, cleaned and transported. Nurses had functional salivary glands, small crops and poison glands, and well-developed ovarioles and Dufour's glands. The third category was that of receivers, which made up the resource sink referred to above. They included the queen, who was fed by foragers; large larvae fed preferentially by nurses but also sometimes fed by some foragers; small larvae fed only by nurses; and callow workers under 5 days old. Lenoir was then able to calculate the resource flux to different groups in each colony, as the percentage of the total duration of food exchanges from foragers. Colonies were rather variable; the queen received 1–50% of the total flux; the larvae 20–80% (distributed about 30% to small larvae and 50% to large ones). Brian (1974) and Brian and Abbott (1977) got similar results with *Myrmica rubra*: nurses took prey juice or syrup from foragers, nurses sometimes passed syrup to foragers but never gave them prey juices. An intermediate group of workers were particularly involved in feeding large sexual larvae. The presence of larvae, and their ability to absorb large amounts of prey juice, caused the flow of prey juice through nurses to larvae, and increased the collecting activity of foragers. In *Solenopsis invicta* large larvae were fed preferentially, and were the only stage to which solid insect flesh was given (Petralia and Vinson, 1978).

In general workers recognize larvae by widespread cuticular substances which can be extracted with fat solvents; the shape of the larva and exudates from its mouth or anus are unimportant, though the cuticular hairs and spines do contribute to recognition. Since these often differ in shape in different larval stages, they may also act as indicators of the age of a larva and help to decide how workers will treat it (Brian, 1975).

A system of resource flow based merely on the priorities that have just been described would channel food to the queen, promoting egg production, and to large larvae, promoting their growth to a size when they might become sexuals. If this were to lead to overproduction of eggs and young larvae it is probable that these would be fed to older larvae, thus reclaiming excess resource from the queen 'sink'. Overproduction of sexual larvae is dealt with in a more complex way. In a series of studies of *Myrmica rubra*, Brian (1983) has shown that sexual larvae are produced from larvae which grow sufficiently fast to reach a certain minimum size before winter. These large larvae do not pupate in autumn but enter a larval diapause. In spring the overwintered larvae have the options of pupating without further growth and becoming workers or continuing their rapid growth and becoming sexuals. Presumably their large size earns them preferential feeding. The inherent instability of a process in which large larvae are fed preferentially and become even larger is counteracted by special behaviour shown by some of the workers related to a special signal emitted by large larvae. Brian (1974) showed that in *Myrmica rubra* workers intermediate in behaviour between nurses and foragers were mainly responsible for the care of these large larvae. Their treatment of them depended critically on the presence or absence of a queen. When small groups were cultured with a queen these workers attacked the large larvae, biting them so as to leave scars on the cuticle; if there was no queen, however, this aggression on large larvae ceased and the larvae grew rapidly. The attacks on large larvae inhibited their further growth and promoted the production of workers (Brian, 1973). The attacks were apparently provoked by a special pheromone produced by a specialized area of cuticle on the ventral surface of large larvae.

3.4 Nest construction

After the production of workers and sexual brood, nest-construction is probably the major area of the colony's efforts, and a considerable amount of time, energy and material goes into it. Since all of these might have been put directly into raising brood, there must be appreciable indirect benefits to brood-raising from the possession of a nest. We can think of nest-construction as investment in the infrastructure of the colony, providing essential services, especially environmental ones like the maintenance of favourable temperatures and humidities for the brood. There is quite a close analogy between investment in housing, roads, power services and sewers in a human economy, and the production of nests, trails, shelters for aphid groups and refuse middens in an ant society. This kind of investment is sometimes referred to as the physical infrastructure. Investment in, for instance, a health service is called the social infrastructure, and this in turn could be compared to the efforts of the nurse-worker force in hygiene and general brood-care. One of the important features of investment in nest structure is that energy so invested cannot, as a rule, be drawn back again into other areas of investment. A colony that has overinvested in brood production, perhaps by the production of major worker or sexual larvae, can solve energy-flow problems by eating these larvae and re-using the material, and some of the energy, for another purpose. Nest structures cannot, however, be re-utilized in this way; ants as a rule cannot re-use materials for further nest-building as bees do with their wax structures. This makes the problems of the economics of construction more severe.

3.4.1 Benefits of the nest structure

The foraging method of ants is a case of 'central place foraging', in which everything they capture is brought back to be fed to the brood. The alternative of moving the brood to the source of supply would produce serious difficulties since the resources on which the colony depends are widely dispersed. Polydomous ants (whose colonies have more than one nest) and ants like *Oecophylla longinoda* have in fact solved this problem to some extent, but the majority of ants retain a central base where food is brought, eggs are produced by the queen(s) and the larvae are raised to maturity. Even the nomadic Dorylinae, like *Eciton*, which do not build a material nest structure and have reduced problems of climatic control in their equatorial forest habitat, retain a daytime base for queen and brood, as well as a night-time bivouc. A second benefit of the nest is the concentration of the queen and brood in a defensible centre; a diffuse brood would have a larger perimeter, and need more guards. This must be as important for *Eciton* as for other ants, and the condensed mass of workers balled round the queen and brood has obvious defensive possibilities. All hymenopteran Apocrita have relatively helpless larvae, which are neither fierce nor mobile and need parental care. In particular the climatic requirements of adult and larva are quite different, and the larvae would soon desiccate in the open. The nests of ants are places of high humidity and have a more favourable range of temperatures than the surrounding air or even the soil. The architecture of the nest also plays an important part in the organization of brood-rearing. Unlike the nests of honeybees or wasps, ants' nests are not cellular in the sense of having separate cubicles for individual larvae. Instead the eggs and larvae are maintained in clumps or piles. This may have been an important factor in the evolution of ant sociality (Malyshev, 1968; see section 2.1.1). Ants' nests are organized in chambers big enough to hold the collections of brood and the workers that tend them. In most species the brood is sorted according to size into at least three categories: eggs and newly hatched larvae, growing larvae, pupating larvae and pupae. These are often found in different chambers, and, since they often have different climatic needs, in different parts of the nest. All these benefits of a nest are achieved in many ants simply by digging into the soil; some colonies, however, improve their nest quality or adapt to special needs by digging under a stone, or by making special structures above ground.

3.4.2 Nest-excavation techniques

Ants probably inherited basic digging techniques from their pre-ant ancestors. If these were soil-nesting tiphioids, as Malyshev (1968) suggests, they would have had digging behaviour similar to that of *Methocha* (Wilson and Farish, 1973). Ants from a wide range of subfamilies dig in soil by pulling out soil particles with their mandibles, and carrying a load of loosened soil to another place where they deposit it. In fine or dry soil the ant may use its forelegs to collect a load together, but tunnels seem always to be cut primarily with the mandibles. The same basic techniques can be used in rotten wood or humus, and in more solid wood (Sudd, 1969; 1982). Additional techniques are used by some species: for instance *Pogonomyrmex* is said to stridulate and so loosen packed dry soil by vibration (Spangler, 1973). The architecture of the nest consists of spaces where soil has been removed, and solid blocks where soil has been left. The actual form of nests in the wild is not easy to study, though a number of attempts have been made either by simple digging or by pouring various casting materials into the nest (Brian and Downing, 1958; Ettershank, 1971; Nielsen and Jensen, 1975). A nest is usually organized into chambers at various depths, connected to each other and possibly to the surface by tunnels. Sudd (1975) tried to derive the structure of nests from the orientations of individual ants digging in the laboratory. Isolated worker ants dug in moist sand, and the shape of the tunnels they produced was often characteristic of their species. Some of the differences between species could be explained by differences in orientation to gravity, or a greater tendency to follow the surface of stones or tree-roots. No chambers were ever produced by isolated worker ants in these experiments, perhaps because immobile workers or larvae are needed to fix the position of chambers.

A further complication of ant digging in real life is that soil is often deposited in tunnels that have previously been excavated, and that the whole nest is often constructed in soil which has already been worked over. Hubbard (1974) showed that *Solenopsis invicta* preferred to dig in soil from its own nest rather than unused soil or soil from a foreign nest; this may be a common mechanism for keeping excavation as compact as possible, and thus more defensible or more thermally efficient. Perhaps it is also more economical to dig in soil that has already been worked. Some of this worked material is carried to the surface where it may be formed into a crater surrounding the nest entrance, or into a mound in which new chambers and tunnels are constructed.

A nest built with these relatively simple techniques may thus become quite a complex structure: chambers at different depths have different temperature and humidity regimes at different times of the day or year. Ants in deserts and steppes often build deeper nests than ants in less severe environments, and winter nests may be at a greater depth than summer ones (Sudd, 1982). *Formica yessensis* in Japan has a nest with superficial chambers and deep vertical shafts. In winter the ants retreat to the shafts, often collecting in oval enlargements (Imamura, 1974). They also move from nests with mounds and large brood chambers in summer to specially constructed ones with deep shafts and smaller chambers in winter (Ito, 1973).

The work done in bringing soil to the surface can be calculated approximately. Sudd (1969) showed that a moderate-sized worker (say about 5 mg in live weight) was capable of excavating about 1 g of moist sand in 24 h. To bring this from a depth of 50 mm would entail 0.00005 J, equivalent to perhaps 10% of the honeydew collected in a day by one ant. In the field *Paltothyreus tarsatus* raises about $200 \text{ g m}^{-2} \text{y}^{-1}$ (Sabiti, 1980), making a considerable contribution to soil-forming processes in African savannah. If this was raised from an average depth of 100 mm, The work done would be about $2 \text{ J m}^{-2} \text{y}^{-1}$, surely a significant proportion of the consumption of a secondary consumer like this ant. The nest-mound of *Lasius flavus* with a radius of say 0.3 m could represent about 50 kg of soil raised say 0.1 m or 50 J. The total production is about $418 \text{ J ant}^{-1} \text{y}^{-1}$ (Nielsen and Jensen, 1975), and perhaps it is significant that mound nests are built by ants that have such large honeydew resources. Even so, the expenditure must pay off in some way by increased security or improved environment.

3.4.3 *Above-ground nest structures*

Even a nest which was merely an excavated burrow would have to dispose of the excavated soil, so that it could not fall, blow or wash back into the excavation. Most ants in fact apply some kind of above-ground treatment to the structure as well as the excavation below ground. The more extensive the underground workings the larger the amount of spoil to be dispersed, and it has been argued (Sudd, 1982) that the conical rings of soil ('craters') around the nests of many ants in warm dry regions may simply represent the most economical way of disposing of the soil safely. If soil was carried the minimum distance which allowed it to be piled at its maximum angle of rest, the result would be a circular rampart of triangular cross-section, whose height and width would depend only on the angle of rest and the volume to be disposed (Figure 3.4). This minimal explanation would not,

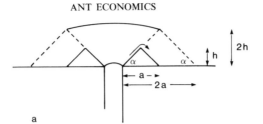

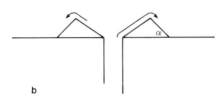

Figure 3.4 The crater nest as the economic solution to soil disposal. (a) A given volume V of material with angle of rest α can be placed in a ring of triangular section and height h and width a, where:

$$a = [V/(2\pi\tan\alpha)]^{1/3}$$
$$h = [V\tan^2\alpha/2\pi]^{1/3}$$

Height and width keep the same ratio $2h/a = \tan\alpha$, and a crater of height $2h$ and width $2a$ holds 8 times the volume. (b) In practice, the inner surface is disturbed by ants and has a lower angle of rest less than α.

however, account for the extra work involved in building crescentic craters, in which earth is piled on only one side of the nest exit, by for instance *Cataglyphis bicolor*. In this species each worker has its own preferred direction, based on a visual orientation, for carrying away soil (Wehner, 1970). Tschinkel and Bhatkar (1974) have shown that in *Trachymyrmex septentrionalis* the crescent is orientated to the general slope of the ground, so that the partial rampart deflects flash floods on the desert surface. This certainly is achieved by the robust circular ramparts of *Myrmecocystus mexicanus*, but why this species should build stronger mounds than *M. depilis* in the same area is not clear. Circular or crescentic ramparts or craters, are typical of ants in tropical and warm arid regions. In cooler climates mounds which cover the nest entrance are more usual. *Cataglyphis bicolor* builds crescents in warm deserts but mounds in the cold deserts of Afghanistan (Schneider, 1971).

Mounds built with earth excavated from below, like those of *Lasius flavus*

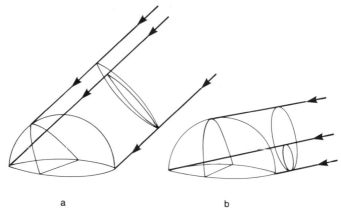

Figure 3.5 Insolation of a mound nest at the equinox in the latitude of Berne. (*a*) At noon a mound receives 1.2 times the radiation as a flat disc of the same diameter; (*b*) at 0700h it receives three times as much. (After Sudd, 1969.)

in Britain, contain the summer nest chambers of the colony. The mound intercepts a greater amount of sunlight than a flat area with the same ground area, particularly when the sun is low. As a result the nest warms more rapidly in the morning and remains in positive heat balance longer in the evening, so that the working day is lengthened (Figure 3.5). This is also more effective at higher latitudes, which may explain why few tropical ants build mounds of this kind; various fire ants (e.g. *Solenopsis invicta*) are exceptions—their mounds may have other functions. Mounds of ants in cool countries vary very much in their strength; *Lasius niger* and *Myrmica rubra* build rather coarse mounds round grass-roots in late summer, and move their brood into the warmer climate this provides. The mounds often do not survive winter, as they are easily destroyed by rain or trampling cattle. *Lasius flavus*, on the other hand, builds mounds which may survive for decades or even centuries (King, 1977). Their durability seems to stem from much slower building around the stems of small herbs, which are able to grow through the deposited earth and bind it together (Brian, 1977).

3.4.4 *Import of special materials for nests*

While it can be argued that the above-ground mounds or craters of ants are merely attempts to make use of an otherwise useless by-product of underground excavation, this cannot be true of ants which transport material to the nest from the surrounding surface. The Australian meat ant

(*Iridomyrmex purpureus*) carries in pebbles which it uses to 'decorate' its nest. This may mark a defended area, like the 'midden' of seed husks and pebbles of the American *Pogonomyrmex*. Ettershank (1971) has put forward evidence that the stones increase heat loss to the surrounding air, or even that the stones are able to condense dew from the air at night. In a hot arid habitat this could justify the outlay of energy in carrying inedible objects to the nest. The outlay can be reduced by using lighter materials to build the mound. This is the case of the mounds of leaf litter built by wood ants (*Formica rufa* group). The density of the mound in *F. polyctena* is only 0.21, about a tenth of the density of rock or soil (Wisniewski, 1967). Brandt (1980) has argued convincingly that in spite of its low density and low heat capacity the materials of the mound of a wood-ant colony simulate the properties of a rock because of its high thermal diffusivity. Absorption of solar radiation is high, and this results in high surface temperatures for a smaller outlay on transport. The temperature gradients which this causes are able to force heat into the mound because of its peculiar structure. The structure is maintained by the constant activity of ants at the surface, which separates an outer layer of pine needles, resin and buds from an inner sponge which may contain as much as 60% twigs (Wisniewski, 1967; Chauvin, 1960). The mound can have temperatures 5° higher than the surrounding soil. Wood ants are unlikely to be able to increase the nest temperature by metabolic activity as honeybees do, since they lack the enormous wing muscles of bees, and can neither 'shiver' effectively nor fan air through their nest. Instead it seems they use energy to construct a more efficient solar nest. If the nest overheats it is cooled by opening ventilation holes in the dome and not by forced ventilation as in bees. Both bees and wood ants have generous supplies of energy-rich sugars to spend in these ways. Wood ants cannot however store energy outside their bodies, as in the winter honey stores of bees. It is probably more economic for them to expend energy in making the nest energy-effective.

3.4.5 *Compounding special materials for nests*

As we have seen, many ants collect and use naturally occurring materials, usually soil or litter, to build their nests. In most cases the materials are not much altered, in contrast to the processing of wood-fibres by social wasps and the production of wax by social bees. The strength of ant excavations seems to depend on careful placing of material, and there is no evidence that they add any bonding materials to their work. Ants which build arboreal nests are rather different, and there are two ways in which they can make arboreal nests out of special materials; by making a strong substance called carton, or by using larval silk to bond leaves and twigs. Some ants are able to nest in pre-existing hollows in rotting trees or in insect tunnels, while others only inhabit trees which provide this facility for mutualists (see Chapter 7). A few species, notably the carpenter ants (large species of *Camponotus*) can excavate in the softer summer wood of healthy trees. These natural cavities are evidently not always suitable for nests, and species like *Leptothorax acervorum* and carpenter ants often build loose partitions from wood fragments inside the cavity. *Lasius fuliginosus*, however, makes partitions inside the cavity with carton. Maschwitz and Hölldobler (1970) showed that carton is an elaborate composite material made of wood (or sometimes soil) impregnated with honeydew and infiltrated by the fungus *Cladosporium myrmecophilum*. Secretions of the mandibular and other head glands may contaminate the carton but contribute nothing to its properties. The ants will only build in the dark, but were studied in a laboratory nest under red light. Three distinct groups of ants were involved. Group 1 ants gnawed wood into pieces about 1 mm across and carried them to another part of the nest. Group 2, which were all foragers, collected sugar solution and regurgitated it to a third group of nurse workers. These took material from a heap deposited by the Group 1 workers and carried it to the building site, regurgitating on it as they carried it. At the building site they laid their loads on the upper edge of the new building, smoothing and kneading it and sometimes lifting it and replacing it elsewhere. They also brought old carton from elsewhere and incorporated it in their work; perhaps this inoculates it with the fungus. The builders worked in small groups at several sites, often on a front 40 mm wide. The fastest building rate in the laboratory was about 4 mm in 24 hours. The fungus grows freely in the carton, and is gnawed down to a shiny sculptured surface in occupied nests: in neglected nests it grows to a lawn-like felt. Tropical *Crematogaster* produce prominent carton nests, more often on the outside of the trunk or branches of trees. The method *Crematogaster* uses to produce its carton is unknown, but it probably is similar to that of *L. fuliginosus*. The carton is made in small flakes about 5 mm across, and these are laid slanting outwards and downwards to throw off water (Soulié, 1961).

Other species of ant use their brood's synthetic powers of producing silk. In many ants, though not in the subfamily Myrmicinae which have lost the ability, the mature larva's last act before it pupates is to spin a cocoon of silk. The silk is secreted by labial glands, and consists of protein fibres. The larva swings its head so as to swathe itself completely in silk, but may be assisted in securing the silk by nurse workers. In *Polyrachis simplex* (Ofer,

1969) and *Oecophylla longinoda* the mature larvae does not spin a cocoon like most Formicinae. Instead, larval silk is used as a nest construction material. In *P. simplex* the silk is used to close the rock cavities in which the ant nests. *Oecophylla* uses its silk in essentially the same way to join together living leaves and to seal the spaces between them. The method it uses has been studied by Ledoux and by Chauvin and is summarized by Sudd (1982). Building involves cooperation, not only like that of *Lasius fuliginosus*, between several groups of workers, but also between them and larvae. The leaves which make up the nests gradually wither and die and the ants are continually building new ones. Building begins when a group of workers collects at the tip of a branch. These workers begin to reach out to other leaves or the tip of the same leaf, and when they have got a grip in both

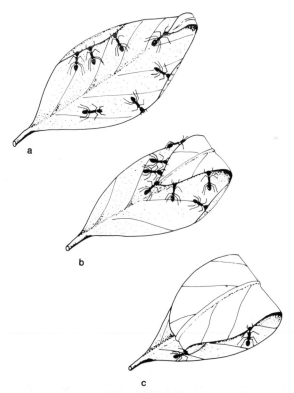

Figure 3.6 The emergence of a 'team' in the construction of a leaf-nest by *Oecophylla longinoda*. (*a*) A few ants begin to bend a leaf; (*b*) ants accumulate where the leaf is bent; (*c*) the bent leaf is secured.

places they begin to pull the leaves together (Figure 3.6). Once an ant has stretched across the gap it is joined by others and a 'group' forms at this point. The strength of several ants is needed to bend the leaves together and to hold them there. Quite often chains of workers form and stretch across gaps that are too wide for a single ant. This phase of drawing leaves together lasts up to 3 h. At about the time the leaves touch, large workers appear carrying half-grown larvae in their jaws. They are carried in an unusual way with their backs towards the carrier's head and their jaws projecting, the reverse of the way they are carried inside the nests. The carriers take their larvae to the leaf margin and hold them near the leaf on a part which is clear of pulling ants. The larva stretches out its head and touches the leaf. Immediately the worker moves it to another point and a thread of silk is drawn out. The worker moves it larva back and forth over the join in the leaves until a zig-zag of silk is formed, and this is closely set so that adjacent threads stick together to make a continuous sheet covering about 10 mm between the leaves. A colony of *Oecophylla*, which has only a single queen, is dispersed into a large number of small leaf nests, any of which may contain larvae, as well as workers and the Homoptera on which they feed.

Crematogaster, *L. fuliginosus* and *Oecophylla* all use nutritious, energy-containing substances to build their nests. So far as we know they are unable to reclaim the substances for recycling: old nests of all three species seem to resist decay and can be found in the wild. The use of such valuable materials is not unique in social insects, for the honeybee builds in wax which it derives from the nectar it collects; but the wax can be reworked for new building. The economics of using such materials has not been worked out, but it must be seen against the cost of the alternatives. The advantages of nesting in the tree where the ant will feed and can dominate have already been pointed out. The cost of carrying soil up the tree would be high, and workers would have to leave the dominated area to collect it. This could entail worker losses or the extension of the dominated area. Again it is probably significant that only ants with very well-developed energy resources, involving the production of honeydew by mutualistic Homoptera, seem to be able to afford to use materials in this way.

CHAPTER FOUR

WHO DOES WHAT, AND WHEN?

One of the secrets of the success of ants, compared with solitary insects that also have nests and show some parental care, is that different individuals in ant nests can specialize in different tasks so that each can work more efficiently. If an ant colony resembles a modern factory (see Chapter 3), a solitary wasp's nest is more like a cottage industry, since the female wasp must alone accomplish a whole series of sequential tasks to raise her offspring. The advantage of a factory over a cottage industry stems from the creation of efficient production lines in which each individual has a small number of repetitive tasks. Similarly, in almost all types of ant colony workers are behaviourally and sometimes anatomically specialized for particular roles such as brood nurses, nest defenders or foragers. Furthermore, by setting up many parallel production lines an ant colony can ensure that if one fails there is a high probability that other lines of production will succeed.

A description of this division of labour in ants must answer the following questions: (i) How do colonies determine which workers are required for which tasks? (ii) How does this task allocation change through the life of the colony and the life of individual workers? (iii) What is the effect of having workers that are behaviourally or anatomically specially adapted to certain tasks? These questions beg a further one: is the organization of the division of labour in ant colonies fully regimented and preprogrammed, or do colonies have the capacity for variable responses in the design of their labour force when unpredictable accidents or crises occur?

Before we can attempt to answer these questions we must determine the kinds of jobs that ant workers need to perform to ensure the survival and growth of their society.

4.1 How ants are employed: how many tasks are performed in ant colonies?

In behavioural studies it is important to have precise definitions to ensure that meaningful comparisons can be made between studies of different individuals and indeed different societies. So we should first be clear about terms: a *task*, for these purposes, can be defined as a set of behavioural acts which achieve some function for the colony, and a *behavioural act* is recognized as a logical unit like grooming, trophallaxis, or carrying a larva. A typical list of such behaviours is presented in Table 4.1. Each task may be made up of one or more behavioural acts.

Table 4.1 An ethogram for *Leptothorax curvispinosus* workers. The relative frequencies of 1962 behavioural acts are tabulated (from Wilson and Fagen, 1974).

Behaviour	*L. curvispinosus* $N = 1962$
1. Self-grooming	23.70
2. Antennal tipping	1.22
3. Groom worker	1.22
4. Groom queen	0.25
5. Carry egg	1.53
6. Lick egg	2.55
7. Carry larva	12.64
8. Lick larva	18.04
9. Help larva moult	0.56
10. Feed larva solids	3.36
11. Carry pupa	1.22
12. Lick pupa	4.84
13. Help eclosion	0.82
14. Lay egg	0.25
15. Regurgitate to larva	7.75
16. Regurgitate with worker	6.42
17. Regurgitate with queen	1.38
18. Fight queen/workers	0.92
19. Lick nest wall	1.38
20. Forage	2.91
21. Eat honey	0.56
22. Eat solids	1.73
23. Carry prey	0.25
24. Carry dead nestmate	0.25
25. Carry live nestmate	0.15
27. Stridulate	0.61
Totals	100.00

To determine how many different behavioural acts occur in ant colonies an ethogram is constructed by closely observing the individuals within a colony and scoring the frequencies at which different behavioural acts are seen. However, even after many hours of observation some rare acts may not have been observed. The problem of discovering rare behaviours in an ant colony is similar to that of discovering all the rare species in an ecological community. Not surprisingly, one can use similar statistics to determine the likelihood that all species or all behaviours have been discovered.

The total behavioural repertoire, i.e. a complete catalogue of all types of behavioural act, can be estimated from a surprisingly small number of hours of observation of ant colonies. In a study of the tiny acorn ant *Leptothorax curvispinosus*, Wilson and Fagen (1974) recorded 1962 separate acts in only 51 hours of observation (Table 4.1). Using a statistic for estimating total behavioural repertoire size they concluded that the 27 different behavioural acts they had observed represented 99.95% of all the behaviours likely to occur in this species. Such 99.95% coverage means that there is only one chance in 2000 that the next act to be observed will be a novel one. By comparison, an equally good estimate of repertory size in a primate or carnivore might require a sample size of 50 000 acts involving 1000 or more hours of observational work. The reason that one can analyse ant behaviour so quickly and accurately is that individuals behave in a relatively simple and stereotyped manner, with a smaller number of rare behaviours than vertebrates.

Total repertoires have been estimated as 28 distinct behaviours in the smallest caste of worker in *Pheidole dentata* (Wilson, 1976a), *Formica perpilosa* (Brandao, 1979) and *Camponotus sericeiventris* (see Calabi *et al.*, 1983), whereas minor workers of *Pheidole hortensis* have an estimated 26 (Calabi *et al.*, 1983). Such minor workers are generally engaged in much the same range of household tasks as the monomorphic workers of *Leptothorax curvispinosus* which also have approximately 27 distinct behaviours. When colonies also possess a physically distinct large caste, of so-called majors, these typically have fewer distinct behavioural acts than their minor sisters. Majors in the turtle ant *Zacryptocerus varians* have 11 (Wilson, 1976b) and majors of *Pheidole dentata* (Wilson, 1976a) and *Pheidole hortensis* (Calabi *et al.*, 1983) nine and six respectively. The relative number of behaviours shown by minors and majors varies in different species. For example, *Solenopsis geminata* minors have 17 behaviours and their majors only two (Wilson, 1978), but in *Orectognanthus versicolor* minors and majors have 27 and 24 respectively (Carlin, 1982). The ant with the largest

Table 4.2 Ethograms for majors and minors of *Pheidole hortensis*. The table shows the frequency of each act relative to the total number of behaviours performed by a physical caste. From Calabi *et al.* (1983).

	Minor	Major		Minor	Major
Selfgroom	.21	.43	Trophallaxis w. major	.001	0
Allogroom minor	.04	0	Trophallaxis w. queen	.001	0
Allogroom major	.02	0	Retrieve food	.05	0
Allogroom queen	.002	0	Forage	.01	0
Carry brood	.45	0	Eat brood/exuvia/dead adult	.02	0
Groom brood	.13	0	Carry waste/nest material	.011	0
Assist larval eclosion	.03	0	Eat solid food in nest	.001	0
Assist pupal eclosion	.001	0	Patrol at food	0	.20
Trophallaxis w. larva	.01	0	Patrol arena	0	.08
Trophallaxis w. minor	.01	.03	Guard	0	.25

TOTALS Minor 1.0, $N = 3697$; majors 1.0, $N = 255$

repertoire so far discovered is the turtle ant *Zacryptocerus varians*, which has been estimated to show between 40 and 42 distinct behaviours (Wilson, 1976*b*; Cole, 1980).

Cole (1980) has demonstrated beautifully that the behavioural repertoires of ants reflect the social design of their colonies and in turn their ecological role. He studied two extremely distantly related types of mangrove ant, the myrmicine *Zacryptocerus varians* and a species of the formicine, *Camponotus* (*Colobopsis*). Convergent evolution of these sympatric species takes three forms: (i) morphological and physiological—for example, both species have majors with shield-shaped heads and the queens produce very large eggs; (ii) qualitative similarities in behaviour patterns, such as very high rates of trophallaxis; (iii) repertoire convergence—there is a remarkable similarity in the number of distinct behavioural acts shown by their workers. Furthermore, when their behavioural repertoires are compared statistically with other species that also nest in mangroves but have rather different ecological roles, *Zacryptocerus* and *Camponotus* are found to be more similar to one another than to any other species.

With this preliminary knowledge of what ants do, we can now proceed to find out who does what. How are these different behaviours and tasks divided up among the workforce? Although we have already indicated that majors appear to be greater specialists than minors, we consider first those ants without any physical distinctions between members of their workforce. There are two good reasons for this choice. First, most ant species have only a single caste of worker: indeed, only 44 of the 263 extant ant genera contain species with more than one physical caste (Oster and Wilson, 1978). Second, as we shall see, the division of labour in monomorphic worker populations is generally based on different individuals having different roles at different times in their lives. Such age-biased differences in behaviour are called *temporal polyethism*. This also occurs in polymorphic colonies; in other words, even majors may do different things at different ages.

4.2 Temporal polyethism: production lines based on an age-biased division of labour

One of the best-known forms of a division of labour based on ageing is that foragers are almost always the oldest workers. There is a very good reason for this: foraging is the most dangerous task; outside the safety of the nest workers run the risk of meeting predators, falling into deadly territorial battles with rival colonies or simply getting lost. Foraging is best performed, from the colony's point of view, by workers that are already close to the end of their physiological lives. A colony should not risk its young workers in foraging expeditions when they could have weeks, months or even years of useful life ahead of them. The limited lifespan of foragers has been clearly shown in one study of harvester ants. Porter and Jorgensen (1981) showed that the foragers of *Pogonomyrmex owyheei* survived on average only 15 days. Indeed, they appear to have been stripped to the bare essentials before they begin this last short phase of their lives. As interior workers progress towards becoming foragers their dry weight declines by 40%. Such foragers can therefore be regarded as a disposable caste. An even more striking example of workers being on Death Row is seen in the Australian weaver ant, *Oecophylla smaragdina*, where the oldest workers occupy special 'barrack nests' around the periphery of the colony's territory. Such workers are the first to become involved in potentially deadly territorial battles with neighbouring *Oecophylla* colonies (Hölldobler, 1983).

The corollary of using the oldest workers as foragers is that the younger workers tend to stay in the nest and occupy themselves with tending the queen, brood care and other household duties. Wilson (1976a) has shown that temporal polyethism also plays an important part in the division of labour among these household tasks. In *Pheidole dentata*, as in many other insects, workers become progressively darker as they age. In *P. dentata* the integument of the youngest adults is clear yellow and gradually darkens to deep blackish-brown as the ant grows older. Using this colour change, Wilson (1976a) was able to discriminate between six different age classes of worker in *P. dentata* colonies. He could also estimate how long a worker was likely to remain in a particular age class by keeping cohorts of the six age classes in separate nest boxes and noting when they first resembled the next age group. It was then possible to construct ethograms, using the methods described above, for sets of individuals selected at random from the six different age classes. This part of the study was based on 2331 behavioural acts, observed in one mature colony, during a five-day period. The data were grouped in to a total of 28 categories of behaviour, two of

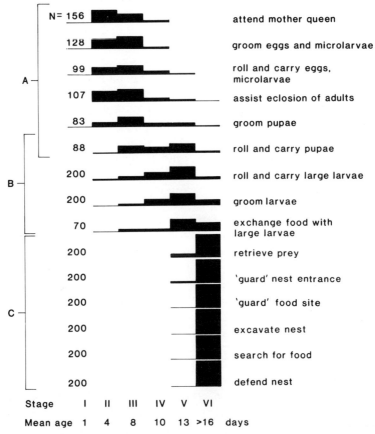

Figure 4.1 The age-based division of labour among workers of *Pheidole dentata*. The proportions of workers in six age groups employed on the principal tasks are depicted as a series of histograms. The total number of observed performances of each task totalled through all age groups are given on the left. The age groups (I-VI) and the average age of workers in each are given at the bottom. The histograms are classified into three groups (*A*, *B*, *C*) which represent three temporal castes. The youngest workers are nurses of the queen and youngest brood, older workers attend old brood, the oldest workers attend dangerous tasks such as nest defence and foraging. (From Wilson, 1976a.)

which, 'guard nest entrance' and 'guard food site' were based on the location of the workers rather than their behaviour as such.

Fifteen of the 28 categories show that the minor workers in this species fall into three largely discrete groups which are called temporal castes, because each group represents a set of specialist labourers whose tasks are decided by their age and therefore vary through time (Figure 4.1). The caste of the youngest workers were engaged in attending the queen, looking after the eggs and smallest larvae, and grooming the youngest adults and pupae. Middle-aged workers formed a caste which tended large larvae, and organized the larval and pupal brood piles. The oldest workers by contrast performed various guard duties and foraged. As a worker ages it moves out from the brood pile to ever more peripheral parts of the colony, and eventually out to its territorial borders. This division of labour is called centrifugal temporal polyethism.

Thirteen of the total of 28 acts considered in this study showed no age bias, or were too rarely observed to be assigned to a certain caste. The behaviours that most clearly showed no age bias included self-grooming, allogrooming (i.e. social grooming) and trophallactic regurgitation with mature workers.

Two of the most important questions about the organization of temporal polyethism are: (i) How many tasks should a certain caste perform? and (ii) How should this task allocation change through the life of a worker? These questions are intimately related to each other and can be answered by considering two opposing hypotheses which can be referred to as a *continuous caste system* and a *discrete caste system*.

4.2.1 Caste systems

To illustrate these opposing possibilities, consider an ant colony with six age classes of worker and four tasks. Model 1 (Figure 4.2a) shows a continuous caste system and indicates the probability that each of the tasks will be performed by any particular age class. In this scheme there is a clear temporal polyethism, as there are distinct changes in the probability that a worker will do each of the four tasks as it ages. However, there is considerable overlap between the tasks allocated to different age groups. For example, a worker in age class 3 will most commonly do task 3, but it may also be found doing any of the other tasks. In other words, the tasks are not grouped into sets with one set for each age class.

By contrast, in a caste system organized in a discrete manner there are also clear changes in task allocation with age but in this scheme certain

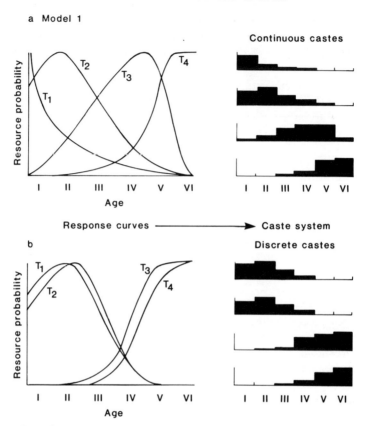

Figure 4.2 A diagrammatic representation of two extreme alternatives in the organization of temporal castes in ants. The age of adult workers is arbitrarily divided into six periods. In model 1 changes in task allegiance ($T1$ through $T4$) change out of phase with one another, so that each of the four tasks is undertaken by a distinct age-class distribution of workers as depicted in the histograms on the right. In this scheme, if there were many more tasks there would be considerable overlap between age-class/task distributions, which would form a clearly continuous system of age-biased task allocation. In model 2, certain age-task associations change in phase with one another to form a discrete system of temporal castes. In this scheme, if the number of tasks was greatly increased the number of age-class/multi-task groups might still remain small. Such a discrete caste system is seen in *Pheidole dentata*, in which the fifteen tasks listed in Figure 4.1 are allocated between only three temporal castes. (From Wilson, 1976a.)

tasks are closely associated. For example, in this Model 2 (Figure 4.2b) tasks 1 and 2 are performed predominantly by the youngest workers and 3 and 4 by the oldest. Moreover, there is a rapid transition between these task pairings as an ant ages; only workers at the transition between age classes 3 and 4 are likely to be found equally employed on all four tasks.

Pheidole dentata is clearly much closer to a discrete caste system than to a continuous one (compare Figures 4.1 and 4.2). Young workers, for example, specialize on those tasks associated with tending the queen and her youngest progeny and helping the eclosion of the youngest adults. Thus young workers in *P. dentata* are the colony's midwives and nurses. When workers such as these are occupied with a distinct set of related tasks, they can be said to have a distinct *role*.

The way this discrete temporal division of labour might contribute to the formation of efficient production lines is if each of the different temporal castes is engaged in a different set of tasks, and each set of tasks is associated with a different part of the nest. Thus when a worker in a certain temporal caste finishes one job it can quickly go on to another within its current repertoire without wasting time and energy moving through the colony: related tasks occur in the same part of the nest. In other words, this organization of castes leads to the greatest possible spatial efficiency within the colony-factory.

However, in another *Pheidole* species, *P. hortensis*, Calabi *et al.* (1983) have described what appears to be a continuous temporal caste system in which task allocation changes through the life of a worker, but there is no association of related tasks and hence there are no distinct roles. Does this mean that *P. hortensis* is losing out on the spatial efficiency that is gained by the discrete role allocation practised by *P. dentata*? Perhaps not: *P. hortensis* has colonies that number a few hundred workers at most whilst *P. dentata* may have a thousand or more workers. *P. hortensis* nests in small discrete sites like nuts whilst *P. dentata*'s large colonies ramify through large logs. Perhaps spatial efficiency is much less important in small factories than in large ones and *P. hortensis* has opted for the advantage of plasticity in role allocation that is not possible with a more precise form of labour and role demarcation. In a sense *P. hortensis* may more closely resemble a family business while *P. dentata* is like a huge corporation.

The importance of colony size as a constraint on caste organization is clearly shown in the primitive ponerine *Amblyopone* which has tiny colonies and is the only known ant with neither distinct temporal nor physical castes in its workforce. The genus *Amblyopone* contains the morphologically and behaviourally most primitive species in the poneroid complex of ants. *Amblyopone pallipes* occurs in woodlands in the northeastern United States and Canada. Colonies can be found under stones, and typically consist of one or more queens and 9–16 workers. Workers forage exclusively underground and are entirely carnivorous, specializing on long thin arthropods, such as centipedes and beetle larvae, which individual foragers drag back to the nest. Both adults and larvae feed directly on the prey and regurgitation is totally absent. Other primitive features include the fact that callow workers emerge unassisted from their cocoons and behave precociously. Traniello (1978) marked all the individuals in a colony with enamel paints so that they could be individually recognized, and constructed simple ethograms for each of the workers.

These data suggest that the activity of *Amblyopone* callows is not directed towards queen or brood care as in other ants. Instead their behaviour is precocious and may involve foraging as soon as five days after eclosion. Temporal castes are apparently completely absent in this species. Perhaps *Amblyopone* colonies are the closest approximation to cottage industries in ants; each worker is a 'jack of all trades' continually responding in a flexible way to the immediate needs of the colony as it meets them. Such flexibility is probably at a premium in small colonies whilst in much more populous ones there will always be, statistically speaking, a predictable number of workers in each age group to perform each essential task.

4.2.2 *Task allocation*

It has been suggested that one efficient and flexible way for a labour force to become organized in small colonies is through task fixation. This would occur if an individual receives a stimulus encouraging similar activity when it successfully performs a certain task. Thus if a worker regurgitates to a larva, is stimulated by this activity and continually encounters hungry larvae there will be an unbroken cycle of positive feedback and it will be effectively assigned to this task. Such task fixation, which involves a special form of learning behaviour, may compensate for the absence of temporal castes in *Amblyopone*, whereas in *Pheidole hortensis* some combination of task fixation, coupled with an age-biased susceptibility to positive feedback from certain tasks, might be the basis of the continuous system of temporal polyethism in this species, which still allows for some flexibility in its work force. This would enable a colony to respond to unpredictable contingencies such as sudden food shortages, changing predation pressures or catastrophic damage to the nest, each of which may cause a sudden loss of specialized workers.

In larger colonies such as *Pheidole dentata* flexibility is at less of a premium because a larger colony population would be better able to absorb emergencies, and so flexibility might be traded for the super-efficiency of regimented production lines associated with discrete temporal polyethism.

However, if a colony's division of labour is based entirely on preprogrammed temporal castes, it must tightly control the age distribution of its workers to maintain its efficiency. This has lead to the hypothesis that the ageing of sterile workers is much more tightly influenced by selection pressures than the schedule of senescence in almost all other types of organism. For this reason, Wilson (1975c; 1985) has suggested that social insect workers may have 'adaptive demographies'.

4.2.3 *Adaptive demography and caste efficiency*

Gerontological theory suggests that organisms that have a last age of reproduction are liable to evolve genes that cause their own destruction. Genes that cause senescence and the physiological death of an individual can still increase in frequency in a population if they also code for increased vitality, survival and fecundity prior to the organism's last age of reproduction. It has even been suggested that the reason many human adults in modern societies survive only to an age of three score years and ten (as was apparently common in biblical times) is due to this form of genetically programmed senescence.

Sterile social insects do not reproduce and do not therefore have a last age of reproduction, and so may be free from this kind of ageing schedule. In societies such as *Pheidole dentata*, workers' ageing schedules also determine their work schedules and their contribution to the productivity of their colony and their own inclusive fitness. Indeed, such adaptive demography could have selected for the extraordinarily long lives of worker ants compared with other insects of similar size (Wilson, 1971). For example, the tiny workers of Leptothoracine acorn ants are known to live for more than two years. *Eciton burchelli* (army ant) workers must survive for 9 months or more, on average, for their colonies to be able to grow at all, such is the relatively low rate at which workers are replaced in the huge colonies of these nomadic carnivores (Franks, 1985).

4.3 Conflicts over the division of labour

These arguments about the efficiency of ant colonies, temporal polyethism, preprogrammed worker careers and adaptive demography are all rooted in the hypothesis that workers are acting, as Darwin (1859) initially suggested, for the profitability of the entire community. However, as discussed in Chapter 1, ant colonies may be rife with conflicts over investment patterns between sexes and over worker or queen production of the colony's male offspring. Indeed, many ant workers are not sterile, and some produce 100%

of the males raised in their colonies (Bourke, 1987). Workers that can produce males may be unwilling to work for the rest of the colony. They will also have a last age of reproduction, and hence may be subject to the type of senescence found in other reproductive organisms. In this case the concept of worker adaptive demography determined by selection on the whole community is likely to be invalid.

Thus reproduction by workers may be one of the chief factors that limits colony efficiency. Reproductive workers may be reluctant to leave the nest, and may actually try to stay close to the colony's egg-rearing chambers so that they can lay their own eggs. As we have already seen, it is the youngest workers who generally tend the eggs and youngest larvae and it is a common observation that such workers generally have the greatest ovarian development. However, even only moderately fecund workers have a last age of reproduction after which they may best serve their own inclusive fitness, and incidentally the productivity of the colony, by altruistically working for the rest of the nest and undertaking such dangerous jobs as foraging and territorial defence. Indeed, the suggestion has been made that such a mixed worker strategy of a period of selfish reproduction followed by a career in altruistic service to the colony may be the evolutionary origin of the temporal polyethism seen in many present-day social insect colonies. West-Eberhard (1979, 1981) has made just such a convincing case for the origins of a temporal division of labour in certain neotropical wasps.

Particularly strong circumstantial support for the hypothesis that the typical centrifugal pattern of temporal polyethism (see section 4.2) began with worker reproduction comes from wood ants, in which the youngest workers have the greatest ovarian development and work in the nest, gradually radiating out from the brood chambers to the nest periphery and later to the foraging systems (Otto, 1958). Such workers probably never produce viable eggs but produce trophic eggs that are a specially rich source of food for other colony members. The ovarian development of these young workers may be indicative of an earlier stage in the evolution of the division of labour in their colonies, when workers did produce males.

One extremely important consequence of reproduction by workers is that it may prevent the evolution of physically distinct worker castes. Individual selection acting on reproductive workers would strongly select for all of them to develop into the single caste that would have the best chances of reproducing. In other words, individual selection on reproductive workers would act against caste proliferation. Oster and Wilson (1978) have examined this possibility by showing that there is an association between (i) the presence of worker ovarian development and monomorphism and (ii) the absence of worker ovaries and polymorphism, in the small number of ant genera and subgenera in which information is available on worker reproductive morphology (see Oster and Wilson, 1978, p. 102).

One of the other major factors that has probably selected for a monomorphic worker caste in all the species of more than 80% of all extant ant genera is the requirement, as summarized earlier, that workers are potentially jacks of all trades and have at least some plasticity in the roles they can accomplish.

However, Herbers and Cunningham (1983) have recently shown that even a typically monomorphic worker population has a division of labour based not only on temporal polyethism but also on individual size.

4.3.1 *A morphological division of labour in monomorphic ants*

The North American acorn ant, *Leptothorax longispinosus*, has monomorphic workers. This means that all the workers are essentially identical in shape; any size variation is small and typically takes the form of a normal distribution (Figure 4.3). These ants are ideally suited to ethogram analyses, as an entire colony can easily be accommodated on the stage of a dissection microscope. Using this technique Herbers and Cunningham (1983) constructed ethograms and time budgets for worker activity. By following randomly selected individuals for 30 min periods they were also able to record the frequency with which one type of behaviour was followed by another, and how workers of different size performed. In this way they were able to construct a table of transition probabilities between behaviours. This indicated the likelihood that certain behaviours would follow others. High transition probabilities indicate that certain sets of behaviours are performed by the same individuals in a short time. If different groups of workers perform separate sets of behaviours they can be said to have distinct roles. In *L. longispinosus*, Herbers and Cunningham documented four distinct roles and three behavioural castes in the workers; one particular caste of workers was responsible for two different roles. In addition there was a significant relationship between a worker's size and her behavioural caste (Figure 4.3). The three castes in order of increasing size were, first, workers performing the two roles of brood care and colony maintenance (the latter being nest-cleaning and provisioning nestmates); second, workers specializing in social interactions; and thirdly, foragers. Thus Herbers and Cunningham conclude that size can underlie polyethism even in an acutely monomorphic species. Ants with weakly dimorphic workers typically have a caste of small workers who act as nurses and a

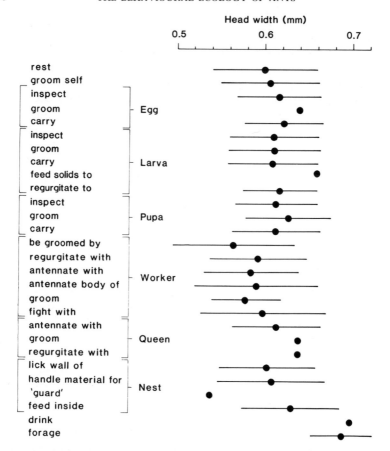

Figure 4.3 Size-based division of labour in 'monomorphic' workers of *Leptothorax longispinosus*. Workers of different sizes, as indicated by head width, tend to perform different tasks. The large dots show the mean head width of workers performing a certain task, the horizontal lines indicate standard deviations. The division of labour in *Leptothorax longispinosus* is probably also partly determined by worker age. (From Herbers and Cunningham, 1983.)

caste of larger workers who act as nest defenders and foragers. Since the undifferentiated small and large workers in *L. longispinosus* have similar roles, Herbers and Cunningham's findings suggest one way in which the behaviour of workers in colonies with distinct physical castes has evolved from monomorphic ancestors.

The other important contribution of this study of *L. longispinosus* was the construction of time budgets as well as ethograms. Clearly the relative importance of different acts and hence role determination should also depend on the relative amount of time workers spend in different activities. Furthermore, time budget studies also serve to demolish a prevalent myth about ant behaviour. Far from being non-stop labourers, whose nests are characterized by ceaseless activity, worker ants spend the vast majority of their time in the nest apparently doing absolutely nothing. A common observation in studies of within-nest behaviour in ants is that workers often spend well over 70% of their total time simply standing perfectly still (Otto, 1958; Cole, 1986). This has led to the suggestion that one of the major roles of ant workers is as reserves in a standing army, alert and ready to come to the colony's defence if and when required. For this reason an ant colony should perhaps not be viewed solely as a factory but rather as a factory inside a fortress (Wilson, 1985).

4.4 Physical castes

Certain species of ants have developed extremely specialized worker castes to perform a small number of unusual tasks. For example, the majors in the myrmicines *Paracryptocerus* and *Zacryptocerus* and in the formicine *Camponotus* (*Colobopsis*) *truncatus* have independently evolved huge wedge-shaped heads that they use to block the entrances of their nests. Majors of *Solenopsis geminata* are specialist millers; their massive heads are equipped with mandibles like huge blunt clubs that they use to grind the seeds that are used by their entire colony for food. In *Pheidole dentata*, the large majors are called into action purely as a specialist defence force against other dangerous or hostile ants, such as *Solenopsis invicta*, which they are quickly able to demolish (Wilson, 1975b). Majors of many *Eciton* species also have disproportionately large heads, but these are equipped with long sabre-sharp mandibles. *Eciton* majors are a specialist defensive caste that serve to protect their colony from would-be vertebrate predators. The ice-tong-like shape of these army ants' jaws can pierce the toughest hide, but the majors do not have the musculature to open them after use. Thus such majors are forced into a kamikaze role, in which they die to protect their colony. When they attack an enemy they remain permanently locked in place, periodically stinging their adversary.

Given the existence of such extreme worker specialization, and of theoretical models suggesting that the greatest possible efficiency would come from one specialist caste per task, it is even more remarkable that distinct physical castes are actually quite rare among ants in general. For example, the British ant fauna, albeit a rather species-poor assemblage, contains not a single species with distinctly polymorphic workers.

The explanation of the rarity of polymorphic worker castes in ants is that there are probably extremely powerful constraints providing counter-selection in favour of monomorphism. We have already considered one such constraint; reproduction by workers which would select for a lack of specialization so that each of them may have, at least, the option of egg-laying. Two other major constraints are (i) economic—specialist castes may be too expensive to produce compared to the rewards they yield; (ii) developmental—given the genetic uniformity of worker populations, physical specialization can only be achieved through a modification of their developmental programme, and only a very limited set of developmental options may be possible. There may not be enough existing or possible variation in worker morphs for selection to act on colonies by picking out castes of a suitable design to perform the required tasks. Both developmental and economic constraints probably play an important part in caste evolution in ants. These constraints will now be considered in turn in the remaining sections of this chapter.

4.4.1 *Allometry as a developmental constraint on caste evolution*

In all known cases, simple allometric growth is responsible for the physically distinct castes that occur among worker nestmates in ants (Wilson, 1953). Allometric growth occurs when two different body parts grow at different exponential rates, so that if the growth of a series of individuals is terminated at different times then the resulting adults would have not only different sizes but also distinctly different shapes.

Consider, for example, allometric growth of the head with respect to the thorax: denote maximum head width and thorax width by the variables y and x respectively. The allometric relationship between these parts could then be expressed as:

$$y = bx^a, \quad \text{i.e.} \quad \log y = \log b + a \log x.$$

Thus a log-log plot of head width against thorax width for a series of such ants would yield a scatter plot in which the variation would fall along a straight line. If the slope (a) of this line is significantly different from unity then the workers at either end of the size range would have recognizably different shapes because the ratio y/x changes as size increases. If a is significantly greater than unity, larger workers have disproportionately large heads. This is how ants form a major caste (see Figure 4.4). Allometric relationships are sufficient to explain the vast majority of the morphological variation in any nest series of worker ants. Moreover, the relative size of

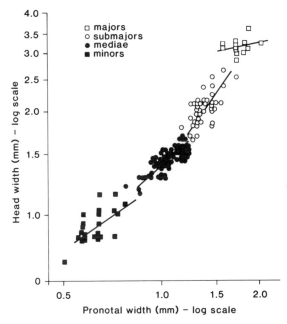

Figure 4.4 Head width/pronotum width allometry of a sample of 150 *Eciton burchelli* army ant workers. Note the logarithmic scales of the axes. The workers were allocated to the four distinct physical castes before measurements were taken. Crosses indicate minims, triangles medium workers, squares submajors and circles majors (see Figure 4.5 for drawings of the heads of these castes). Straight lines representing the allometric curve for each caste have been fitted to the data by the least squares method. (From Franks, 1985.)

a particular pair of body parts over an entire series of workers derived from one polymorphic nest population can in most cases be described by just a single allometric curve. In the most extreme cases of polymorphism, however, the curve may have two or three break points, e.g. where further morphological variation has been programmed by switches in the larval development of workers. However, these breaks in the overall allometrical relationship are always relatively small (Figure 4.5).

This suggests that there have been severe limits to the range of worker morphs that can be created in ant colonies in the course of their evolution (Wilson, 1985). Conformity to allometric rules means that a marked change from, say, a minor to a medium, would if continued have a huge influence on the morphology of majors. Thus changes in smaller workers might result in the production of non-viable larger castes and, vice versa. A colony can in the course of evolution dispense with intermediate forms so that distinct

82   THE BEHAVIOURAL ECOLOGY OF ANTS

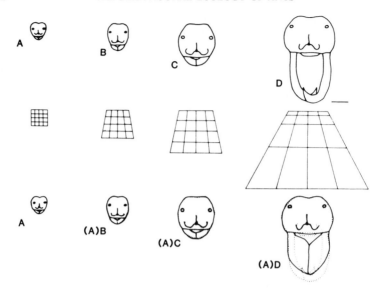

Figure 4.5 The top row A, B, C, D are the heads of *Eciton burchelli* minim, medium worker, submajor and major respectively. The grids below each caste represent the form of the cartesian transformation required to modify the minor caste into each of the larger forms. $A(B)$, $A(C)$, $A(D)$ are the modified minor, the solid line, superimposed on the dotted outline of the respective larger caste. The scale bar to the right of D represents 1 mm. (From Franks and Norris, 1986.)

populations of majors and minor forms can be recognized with few transitional forms. Because of their allometric relation any one caste still cannot be altered without affecting the others.

In *Eciton burchelli*, for example, there are four distinct worker morphs; in order of increasing size these are the minor, medium, submajor and major castes. Majors are created by extending and distorting the growth pattern of submajors which, in turn, are a larger distorted version of medium workers, whereas minims are a smaller distorted form of medium workers (Franks, 1985).

Franks and Norris (1986) recently used computer graphics to model the whole head morphology of different castes of ants in a number of species in different phylogenetic lines. A single transformation rule, in which allometric formulae of the form $y = bx^a$ are used to manipulate two-dimensional line drawings, can model the relative shape of, for example, the head of an *Eciton burchelli* major compared to a submajor. The same transformation will model the relative size of the mandibles, the width of the head, the position of the eyes, and the location of the antennae in these different castes. In other words, the positions of all of the most important organs of the head are determined by the same transformation rule. An *Eciton burchelli* major with relatively longer mandibles would have eyes practically on top of its head (Figures 4.5, 4.6). This implies that in many ant species the process of creating a useful major caste might involve the production of a range of useless intermediate 'monsters'. The cost of these useless forms would be so high that even the potentially useful majors could never evolve.

The restrictions imposed by simple allometric growth are probably the major reason for the low number of physically distinct castes in any one type of ant colony. Only three ant genera of the 263 extant ones have more than two worker castes, and, among these, four distinct physical castes of worker are known in only one species, the army ant *Eciton burchelli*.

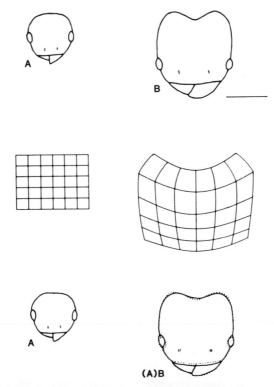

Figure 4.6 *A* and *B* are a *Pheidole* species minor and major respectively. The grids and lower heads follow the format explained in Figure 4.5. The scale to the right of *B* represents 1 mm. (From Franks and Norris, 1986.)

4.5 The economics of caste ratios

Since physically distinct worker castes are common only in species in which workers have little or no reproductive potential, it is necessary to analyse the design of sterile polymorphic workers in terms of their contribution to the productivity of the colony as a whole. In other words, we can examine the divisions of labour among physical castes in ant colonies as if they are the result, over evolutionary time, of a series of economic experiments in colony design. Colonies with the best-designed factory-fortresses, including the most appropriate workforces, should have produced most sexual offspring and therefore most new colonies. An analysis of modern ant colonies should therefore detect patterns of superior design. This approach has been called the study of ant ergonomics, to draw an analogy between studies of the distribution of work, performance and efficiency in human and social insect economies (Oster and Wilson, 1978).

By applying to social insect colonies forms of micro-economic theory and optimization analysis first applied to man-made factories, certain predictions have been made. As a physical caste becomes more and more physically specialized through its evolutionary history, i.e. as it becomes more and more adapted to a single or a restricted number of tasks, then (i) its relative abundance in the colony should decline and (ii) its behavioural repertoire should become ever smaller.

These predictions are born out in several studies of different ant genera. For example, Wilson (1978) examined the behaviour and degree of specialization in the minor and major castes of *Solenopsis invicta*, the imported fire ant of the United States, and the 'native' fire ant *Solenopsis geminata*. The latter is the more advanced species, in terms both of the morphology of its major caste and of its overall division of labour. Both *Solenopsis* species have distinct majors, but in *S. geminata* the majors are relatively larger and rarer (Figure 4.7). In *S. geminata* allometrical differentiation among the workers has produced majors with massive heads equipped with huge blunt mandibles for grinding seeds. Associated with this extreme physical specialization, *S. geminata* majors have a behavioural repertoire consisting of only two acts—the smallest yet recorded for any social insect. By contrast, *S. invincta* is less of a seed-eating specialist; its majors are less morphologically specialized and have a larger repertoire.

Similar trends are seen in the genus *Pheidole*. Wilson (1984) examined caste ratios and the division of labour in 10 species of *Pheidole* from various localities in North and South America, Asia and Africa. In all of these species, minor workers had similar behavioural repertoires but the different

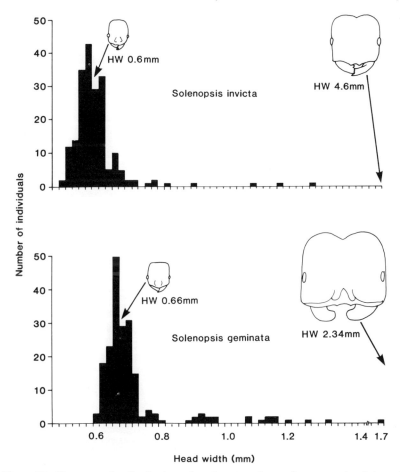

Figure 4.7 Frequency size distributions of worker castes in two fire ant species, *Solenopsis invicta* and *S. geminata*. Both distributions are based on random samples of 200 ants taken from laboratory colonies. Both distributions are skewed to the right, especially that of *S. geminata*, from the normal distributions found in monomorphic species. The outlines shown are minors and majors at the two extremes of the size variation.

majors showed from 4 to 19 distinct behavioural acts depending on the species. Each species had a characteristic proportion of majors in its workforce, and the size of the behavioural repertoire of majors was positively correlated with their relative abundance (see Figure 4.8). Thus if majors represent a higher percentage of the total workforce then they contribute more widely to the general economy of their colony. In these 10

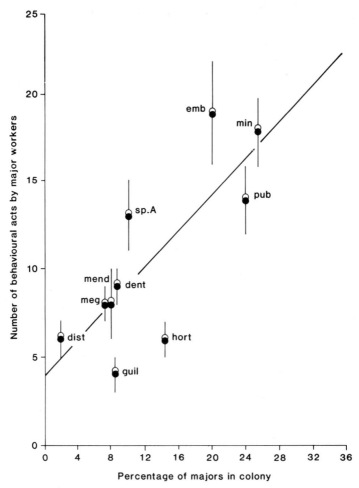

Figure 4.8 The relationship between the percentage of major workers in colonies of 10 species of *Pheidole* (*P. embolopyx, P. minutula, P. pubiventris, P.* (species) *A., P. dentata, P. megacephala, P. mendicula, P. distorta, P. hortensis, P. guilelmimuelleri*) and the size of the behavioural repertoire of the major caste. (From Wilson, 1984.)

Pheidole species there was also a negative correlation between major size relative to their conspecific minors and the size of the behavioural repertoire of the majors. Thus as majors become relatively larger and hence more specialized they become relatively rarer in colony populations and do fewer things. Both of these findings are consistent with the predictions of ergonomic theory.

4.5.1 Time and motion studies

Not only can ergonomic principles predict gross patterns in the abundance, morphology and division of labour of physical castes among different species, they can also reveal why certain castes specialize in certain tasks and explain precise work patterns in social insect colonies.

Such studies of social organization in ants, like physiological investigations, address the wider biological problem of how simple units can become integrated to produce a more complicated and sophisticated whole. Indeed, since the inclusive fitness of the individuals within the most advanced eusocial colonies (such as *Solenopsis*, *Pheidole*, *Atta* and *Eciton*), in which workers cannot reproduce at all, is tied up with the productivity of the whole colony, such colonies can be regarded as self-sustaining, homeostatic systems like a single 'super-organism'. However, there are great technical advantages in analysing how the parts of social insect colonies are put together compared with investigations of the integration of single organisms. A social insect colony can be partly disassembled and re-assembled by an investigator in a way which would be completely impossible with a single complex organism. Castes can be separated and their performance assessed, or caste ratios can be adjusted and then reconstituted in order virtually to rerun the evolutionary history of a certain pattern of social organization. Some of the various options that natural selection has discriminated between and discarded in the past can be recreated as so-called 'pseudomutants' (Wilson, 1980a, b) to determine if evolution has actually resulted in an optimized design for a particular kind of ant colony.

In a particularly beautiful series of studies Wilson (1980a, b; 1983a, b) has used this 'pseudomutant' technique to examine the division of labour in colonies of the leaf-cutter ants, *Atta sexdens* and *A. cephalotes*.

4.5.2 Case study: the ergonomics of leaf-cutter ants

Atta colonies are excellent subjects for ergonomic studies. They are almost entirely devoted to processing living vegetation into a substrate for their fungus gardens. This means that the economics of foraging by a colony can be assessed simply in terms of the amount of vegetation gathered per unit time and the cost of the harvesting procedure. In addition, perhaps more than for any other ant, there is a distinct production line in these colonies: medium-sized workers cut and retrieve leaves, and successively smaller workers process this material until eventually it is in a form into which the fungus can be planted. The very smallest workers tend the growing fungus, as dexterous gardeners, weeding and selecting the fungal tufts hypha by hypha. These tiny minor workers also dispense this fungal food to the rest of the colony. In the leaf-cutter ant *Atta sexdens* size variation is continuous, and all of the size groups together perform a total of 29 tasks. The physical castes cannot be segregated into discontinuous groups on the basis of discrete morphological or anatomical traits. However, four castes can be recognized on the basis of role clusters which are associated with fairly distinct activity curves (Figure 4.9). The four behavioural castes are

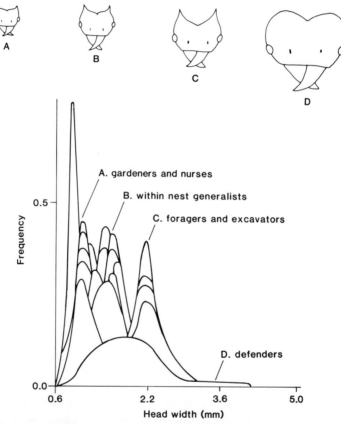

Figure 4.9 Workers of *Atta sexdens* can be segregated into four castes largely based on physical size. The simplified frequency distributions are for workers of different size performing each of 22 tasks. The frequency curves fall into four groups called role clusters. (From Wilson, 1980a.)

gardener–nurses, within-nest generalists, forager–excavators and defenders. To draw attention to this form of coupling of size variation with task specialization, Wilson (1978) has termed this association between physical morphology (allometry) and the division of labour (polyethism) *alloethism*—i.e. the regular change in behaviour patterns as a function of size.

Against this background of the social design of *Atta* colonies, Wilson (1980*a*, *b*) first asked a simple question: Is vegetation harvesting undertaken by the most appropriate size of worker? To test this, workers in a laboratory colony were allocated to size classes in terms of their head widths. (Head width is not only a standard taxonomic measurement, it is also relatively easy to determine even for living ants and is least prone to error as, due to allometry, head widths increase disproportionately rapidly compared with other aspects of larger body size.) Leaf-cutting was selected for ergonomic study because in *Atta sexdens* it is performed largely by workers of medium size (head width 1.8–2.8 mm), who do very little else and should therefore have evolved morphologically and behaviourally for this one task.

Wilson used his pseudomutant technique to assess the relative performance of different size classes of worker in the role of leaf-cutting. Groups of foraging workers were removed until workers of only one size class were left outside the nest. The performance of this remaining group was then assessed in terms of their rate of attraction to the task, initiative in leaf cutting, and rate of leaf retrieval. To convert this measurement of performance into the hard currency of rate of energy acquisition such measurements as oxygen consumption, as a function of body weight and head width, and running velocity were also made. These data were then used to evaluate three competing hypotheses for different ergonomic criteria. One possibility is that the chief selection pressure on foraging would be predation, which the ants might minimize by means of defence and evasion. Another possibility is that the foragers would minimize foraging time through 'skill' and running velocity. Lastly, the chief criterion might be energetic efficiency so that the foraging specialists would be simply the most appropriate workers to harvest energy at the least possible cost. Such costs include not only the amount of energy workers consume during foraging but also the amount of energy that was required for their own construction. For example, smaller workers may be slightly less efficient day-to-day than larger ones, but if they are much cheaper to build and maintain, then natural selection, acting at the colony level, should favour the smaller caste.

Using the pseudomutant techniques and careful measurements of the

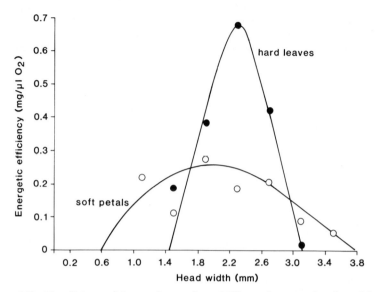

Figure 4.10 The efficiency of *Atta sexdens* workers of different size assessed as the weight of vegetation they harvest per unit time per unit dry body weight (mg) divided by maintenance costs, i.e. their energy consumption while resting at 30° C. Workers with a head width of about 2.2 mm are most efficient, and in nature are those most commonly employed in vegetation cutting. (From Wilson, 1980*b*.)

relevant parameters, Wilson was able to show that the size-frequency distribution of leaf-cutters in *A. sexdens* conformed closely to the optimum predicted by the criterion of energetic efficiency when the ants were utilizing tough vegetation. This distribution is optimal with reference to both construction and maintenance costs (Figure 4.10).

Economic analyses can be applied, of course, not only to large factories at steady state but also to patterns of investment as small firms grow into bigger ones. This kind of analysis is particularly fascinating for the initial stages of colony growth in ants, in which the factory-fortress often begins life as a single queen claustrally forming a nest and relying entirely on the breakdown products of her now useless wing muscles to grow her first workforce (see Chapter 3). As in human endeavours, small new firms are most vulnerable to bankruptcy. For this reason queens and their new colonies should be subject to the most rigorous selection to invest their very limited resources in precisely the most appropriate size classes of workers in relation to unit cost and efficiency. This is exactly what happens in leaf-cutter colonies. Small colonies of *Atta cephalotes* with between 20 and 60 individuals are composed almost entirely of workers with a head-width range of 0.8–1.6 mm. Smaller workers than this are produced in mature colonies, but this size range corresponds to the smallest forms that can both harvest fresh vegetation and also cultivate the symbiotic fungus. Thus in founding the colony the queen produces close to the maximum number of workers that can collectively perform all of the essential tasks.

As the leaf-cutter colony grows through the first 3–4 years of its life, gradually both the biggest and smallest size classes of worker emerge and the size-frequency curve of workers becomes progressively more skewed to the right. However, during all these founding years, the chief investment remains in medium workers with head widths of 1 to 1.6 mm, i.e. the smallest caste that can still accomplish all the vital tasks of colony life (Wilson 1983a, b). Only after many years of ploughing back the profits from the efforts of these and subsequent generations of workers will it start to produce large majors.

4.5.3 Case study: the ergonomics of foraging in army ants

The huge foraging systems of army ant colonies, coupled with the extreme polymorphism of their workforce, provide an unrivalled opportunity for comparative studies of worker ergonomics. For example, the army ant *Eciton burchelli* produces larger raid systems than any other neotropical army ant and has four distinct castes of worker; more than any other known ant species. The minors probably act as brood nurses. Medias are equally represented in all aspects of colony life and are the generalists. Submajors are specialist porters, and the majors are for defence. Table 4.3 shows the relative frequency of these four castes in the bivouac, in newly enclosed cohorts of callows, in the raid system and in the role of prey transport. Certain aspects of this table are particularly noteworthy. First, majors are generally very rare and never carry items; indeed, their ice-tong-like mandibles completely bar them from this role. Second, minors are extremely rare as prey transporters; as shown later, this is probably because they are particularly uneconomical in this role. However, submajors are disproportionately common as prey transporters; they are only 3% of the total workforce, but are 26% of prey porters.

Prey transport is a particularly important aspect of the foraging efficiency of army ants. An *Eciton burchelli* colony may retrieve 30 000 or more prey items in a single raiding day, and must carry these items very considerable distances. The average raid distance at the termination of foraging is 105 m. For the colony, time spent retrieving so many items such

Table 4.3 The relative frequency (%) at which the four physical castes of *Eciton burchelli* workers are employed in different parts or different roles in the foraging army ant colony (From Franks, 1985).

	Minim	Medium	Submajor	Major
Bivouac ($N = 573$)	33.68	62.13	3.14	1.05
Brood ($N = 400$)	29.75	68.00	2.00	0.25
Raid ($N = 3314$)	22.00	74.55	3.15	0.30
Carrying prey ($N = 244$)	7.38	66.39	26.23	0.0

great distances is time lost from pushing forward the raid system and capturing more prey. There are two important aspects of transport efficiency: transport costs and retrieval speed. Transport cost, defined as the energy used per unit weight per unit distance, declines with the increasing size of the carrier. This is true both of terrestrial animals and of man-made machines. It is cheaper for an elephant to move a percentage of its body weight through a given distance than for a mouse to move the same percentage of its body weight a similar distance (Alexander, 1982) just as it is cheaper (per item) to move a load of groceries in a huge truck than in a small van. This justifies the disproportionate employment of submajors in prey transport as these workers are much bigger than the vast majority of their nest mates. The larger majors are barred from prey transport by the design of their mandibles which is a consequence of allometric growth. The decline in transport costs with size means that it could actually be cheaper for a submajor to carry a minim worker than for both of them to travel under their own steam on a long journey. Such portage of nest mates actually occurs quite frequently during army ant emigrations (Rettenmeyer, 1963).

Prey retrieval speed will also be important to army ants as, unlike leaf-cutter colonies that are rooted in place, army ants must complete the raid in a limited period before the entire colony moves to a new nest site. Submajors are not only the most economical porters of prey but can also run faster, when unburdened, than any of their nestmates. This means that they have a short turnround time between delivering one load to the bivouac and returning to the raid front for the next. The speed of submajors may be attributed to their unusual morphology: they have longer legs in proportion to their body size than any other caste. In a whole range of terrestrial animals, leg length is positively correlated with running velocity.

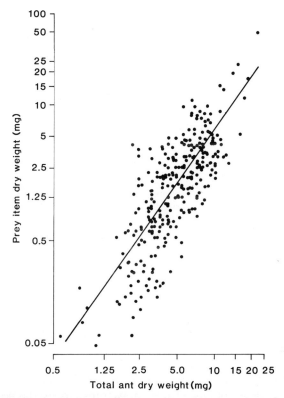

Figure 4.11 The relationship between total army ant worker dry weight (A mg) for both lone porters and groups of porters, and the dry weight of the prey item (I mg) they carry: $I = 0.1208 A^{1.606}$ ($n = 312$, $r = 0.787$, $p < 0.001$). The transition between loners and groups occurs between 5 and 10 mg total ant dry weight. A larger weight of ant or ants can carry disproportionately heavy items. (From Franks, 1986.)

In these ways submajors seem to have been specially designed as a road-haulage caste.

Most remarkably of all, *Eciton burchelli* workers appear to form highly structured and super-efficient teams to retrieve extremely large prey items. *Eciton burchelli* workers, like all army ants, carry items slung underneath their bodies; this enables many workers to cooperate easily when they carry the same item. In such groups each worker straddles the prey item, and they can all run with their bodies orientated in the same direction. In this species, too, larger workers carry disproportionately large items. A plot of log (ant dry weight) against log (prey item dry weight) has a slope significantly greater than unity. In addition, the plot of group dry weight against prey item dry weight follows exactly the same relationship as that for lone porters (Figure 4.11) (Franks, 1986). All prey items are carried at constant speed, probably to avoid traffic jams. This means that group transport is super-efficient. Groups carry items that are so large that if the prey item was fragmented so that there was one piece for each member of the former team, they would be quite unable to carry the fragments. This clearly explains why teams are formed: more prey can be moved with fewer workers. One possible explanation for the super-efficiency of teams is that team members coordinate their leg movements so as to keep more legs on the ground per unit load than is possible for individuals. Thus workers may modify their behaviour when they act as a group to fall out of step with their nestmates.

This extraordinary ability of ant workers to cooperate so that their combined efforts may be more than the sum of the performance of individuals, coupled with the resilience that a workforce of generalists provides against crises during the life of the colony, has probably alleviated selection pressures towards more and more specialist castes in ants.

4.6 Caste ratios and social homeostasis

One of the most exciting and newest areas of investigation into caste ratios and the division of labour in ants is the study of how colonies respond to changes in their environment. How plastic and versatile is the design of a colony's work force? How resilient are colony caste ratios to perturbations? More specifically, are caste ratios governed by control systems that continually reshape the workforce to the changing requirements of the colony?

Colonies in nature, given their long lives, will be relatively commonly subject to catastrophes like the removal of part of their workforce by an influx of predators or a flood wiping out most of their foraging population. There are two different ways in which a colony could respond to such sudden changes in the composition of its workforce. First, the remaining workers might expand their behavioural repertoires to cover the tasks of the missing workers. Second, brood development could be swiftly channelled into the disproportionate production of the missing castes. In fact, both behavioural elasticity and changes in worker production contribute to the resilience of a colony's economy. This has been shown by an experiment in which 3–4-year-old *Atta cephalotes* colonies had their workforces reduced from about 10 000 to less than 250 workers, with a size-frequency distribution typical of young colonies of the same size. The colonies responded by producing a new brood of workers whose size was characteristic of a small colony rather than a large one (Wilson, 1983a, b). This is exactly what one might expect from economic principles. The best strategy for a colony of small size, whether it is old or young, is to use the standard growth trajectory of colony development in order to grow most quickly to a size at which it can produce sexual offspring and realize its inclusive fitness.

However, in the above case the question remains as to whether *Atta* colonies are truly plastic in their response to a crisis or are simply capable of returning to a 'hard-wired' growth program. Evidence that caste ratios are an innate, and largely predetermined, aspect of colony design comes from studies of *Pheidole dentata*. As described earlier in this chapter, *P. dentata* workers appear to be a specialist defensive caste against colony invasion by foreign workers such as those of *Solenopsis invicta*. Colonies of *P. dentata* continually stressed by exposure to *S. invicta* workers for a period of 19 weeks (more than three times the developmental time of major workers from egg to pupa) completely failed to alter their rate of production of majors. Furthermore, different *P. dentata* colonies have very different ratios of minors to majors that appear to be an innate and constant feature of their own social design. When colonies are experimentally perturbed through the removal of parts of their workforce they produce new workers in proportions that re-establish their original unique caste ratios (Johnston and Wilson, 1985). Again, such apparent resilience does not necessarily imply any great plasticity on the part of the colony. One of the best-known principles of population biology is that an age-structured population growing exponentially, with fixed natalities and mortalities per age class, will relatively quickly move towards a stable age distribution. In other words, the proportion of individuals in each class will quickly tend to become constant. Caste ratios in *P. dentata* colonies might be analogous to such age classes, and the constancy in the ratios may be a simple result of the laws of population growth, rather than any plastic resilience.

The one aspect of social design where ant colonies are known to show great resilience through the measured response of their workers occurs through behavioural elasticity. When 90% of the most efficient size class of leaf-cutting workers are removed from large *Atta cephalotes* colonies, the workers from neighbouring size classes quickly fill in for their missing nestmates. The rate of leaf harvesting remains unaffected by such a crisis because sufficient workers in the size classes adjacent to the most efficient (and missing) one are already present in the foraging area, apparently on a stand-by basis. Thus enough of these substitutes are already present in the foraging area and more do not have to be called up from the nest.

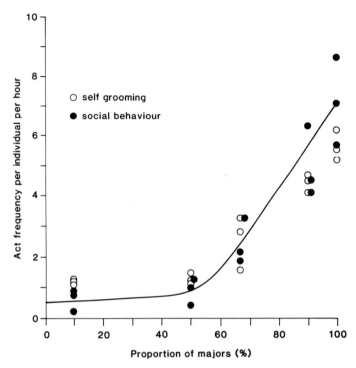

Figure 4.12 Rates of activity in social behaviour and self-grooming as functions of the proportion of majors in artificially-constituted subcultures of the South American ants *Pheidole pubiventris*. Social behaviour is defined as all of the repertory except self-grooming, feeding and individual waste disposal. As minors become rarer, majors become more and more active to replace the labour of their little sisters. (From Wilson, 1983.)

Moreover, the few remaining 'optimal' workers for the task of leaf-cutting stepped up their individual activity by a factor of approximately 5 (Wilson, 1983a).

Similar behavioural elasticity is seen in response to colony crises in a number of *Pheidole* colonies. The usual range of the ratios of the minors to majors is 3:1 to 20:1 according to species. When the ratio was lowered to below 1:1 in three widely different species of *Pheidole*, *P. guilelmimuelleri*, *P. megacephala*, and *P. pubiventris*, the repertoire size of majors increased by a factor of 1.4–4.5 and their rate of activity by a factor of 15–30. Thus the often ungainly majors took over the nest duties of their missing little sisters (see Figure 4.12). Perhaps more remarkable still, the change occurred within 1 h of the imposed change in ratio and was quickly reversed when the minors were restored.

The findings for both *Atta* and *Pheidole* colonies suggest something quite unexpected and wonderful. Workers in a particular caste seem to be fully aware of the presence and performance of their nestmates, and will quickly step in to the gulf if their colleagues' performance is not up to scratch. How could this mutual monitoring be achieved? When Wilson (1984) manipulated caste ratios in *Pheidole* he noticed that the major change in the behaviour of majors, when minor numbers were reduced, was associated with brood care. In *P. pubiventris*, for example, majors rapidly increased both their rate of activity and their behavioural repertoire by starting to care for the colony's brood when minors were removed. When minors were replaced, majors quickly reverted to their normal behaviour patterns and ceased brood care. To determine how these swift changes come about, Wilson (1985) subcultured parts of a large *P. pubiventris* nest so that the colony fragments had a wide range of major/minor ratios and a quantity of brood proportional to that in the original whole colony. By closely comparing the behaviour of majors and minors within 1 mm of brood and at greater distances ($>$ 1 mm) from the brood, Wilson was able to show how majors could be monitoring and responding to the abundance of minor workers in the role of brood care. The very simple response of majors involved avoiding minors when they are near brood items. Majors showed a distinct aversion, scored in the behavioural investigation as the frequency of turns through 90° or more when they encountered a minor near brood. By contrast, minors do not display any form of aversion to either other minors or majors when they are near the brood. In this way minors can quickly take over their old jobs when returned to the nest and will rapidly displace majors in the role of brood care.

In this way *Pheidole* workers could be subject to the counter-balanced forces of task fixation (see Section 4.2), (leading them to perform the most neglected tasks), and differential caste aversion (leading them to leave tasks to the most appropriate of their nest mates if any are available. Thus these ant colonies seem to have great scope for social homeostasis involving measured responses on the part of individual workers to the needs of their colony.

CHAPTER FIVE

COMMUNICATION

As well as their physical resources of workers and sexuals, and environmental resources, ants also deal with a less tangible resource—information. Ant societies do not invest so much in information as the modern global human society does, but they do at very least invest a great deal of time in the collection and diffusion of information. Information probably has in fact two main roles. First, the 'cohesion' of the society requires that its members should 'know' one another, so that the benefits of the society may be confined to its members and so maintained within certain limits of relationship. Only in this way can membership of the society increase individual inclusive fitness. Many signals are therefore concerned with recognizing colony affiliation as an approximation to kinship—there seems to be little evidence that ants can recognize genetic relationship—and with recognizing the queen, and coordinating her reproduction and that of workers. The second role of information is in coordinating the exploitation of environmental resources; in facilitating the discovery, recovery and monopolization of resources, and in the defence of the capital (e.g. brood) and infrastructure (e.g. nests, etc.) in which resources have already been invested.

5.1 Ant signals and language

Early students of ant behaviour were often impressed by the ability of ants to recognize their nestmates, to find their way to their nest and to direct nestmates to newly found food resources. While a hypothetical 'nest-odour' was soon suggested as a physical basis for nestmate recognition, some writers resorted to concepts like telepathy or a 'spirit of the hive' to explain otherwise inexplicable powers of ants. The discoveries of Karl von Frisch between 1928 and 1946, of the dances of the honeybee, and also discoveries of the chemical communications systems it uses, provoked a search for similar communication channels in ants. Since ants are quite frequently observed touching each other with their antennae, much emphasis was at first given to the existence of a postulated antennal language among ants.

More recently, properly planned experiments and the development of chemical methods capable of detecting and identifying very small amounts of chemicals have helped us to realize the importance of chemical channels among ants. Substances used for chemical communication between members of the same species are called *pheromones*. There are two important groups of pheromones in ant societies. The first is the 'colony odour', which seems to reside in the cuticle of all adult members of the society, and is used to discriminate nestmates, to whom the resources of the society may be passed, from aliens who should be attacked. This often also sets the context in which other signals are interpreted; for instance an alarm pheromone may induce attack on aliens but not on nestmates. The second is the range of pheromones produced from specialized exocrine glands. Each of these pheromones elicits specific behavioural responses in members of the same species. For this function chemical signals have one very clear advantage. Provided that an ant can synthesize enough distinct chemicals with reasonable economy, and these chemicals are unlikely to arise in the environment except by the action of ants, the meaning of each signal (i.e. the appropriate response) should be unambiguous. Against this we must set one or two drawbacks. The ant is obliged to synthesize special substances at some cost to itself: naturally-occurring cheap products like carbon dioxide or water have limited communicative value. Although the meaning is unambiguous, modulation of the signal to carry more information is difficult. In particular, since the chemical signal is propagated by diffusion, or worse still by small-scale turbulence, its range and its strength, as actually received by another ant, are not under the control of the sender. The actual speed of propagation depends on the physical properties of the pheromone substance, particularly its volatility and its diffusion constant. As a rough guide, substances suitable from a physical point of view would have molecular weights between 100 and 200. The upper limit on molecular weight sets a limit on the number of different compounds which can be used: the larger the molecule (or in practice the greater the number of carbon atoms), the greater the possible combinations. The dependence on random processes of diffusion and turbulence for propagation means that it is not possible to add further information, for instance by pulse coding, since this would be overwhelmed by the element of 'noise' added during transmission of the message. There are also problems in 'switching off' the message, since the substance is likely to persist after its role is completed. In some non-social insects special 'switch-off' pheromones are secreted to annul the unwanted signal.

Mechanical signals are in general less specific, and have to be received against a background of spurious signals caused by non-communicative movements. In animals living above ground this problem is often eased by the development of bright colouration to mark out parts of the body as organs of communication (for example the coloured palps of wolf spiders and the plumage of birds). This option is not open to ants in many contexts, since the signals have to be transmitted and received in the darkness of the nest or while hunting in soil or vegetation. Communicative movements (signals) must therefore be marked off from insignificant disturbances (noise). One method of doing this is to give the signal an unlikely form either by repeating it in a regular manner (as in the repeated dances of honeybees and the repeated antennal tapping of ants), or alternatively by using a carrier movement of improbable frequency (like stridulation), which is modulated to carry the message. Stridulation, usually by the friction of the node on the gaster, is used in some ants at least for the transmission of messages through the soil. A third possibility would be to accompany the mechanical message with an unambiguous chemical marker which declares it to be significant. Antennal and other movements are an obvious part of ant life, and must be assumed to play a part in communication, whether by providing a context for chemical communication or as a parallel system to the chemical channel.

5.2 Recognition of nestmates

The societies of ants and other social insects are closed societies: it is not usually possible for any insect, even of the same species, to enter the society. Closure is preserved by the need for an 'olfactory visa' (Jaisson, 1984) or 'discriminator' (Carlin and Hölldobler, 1983) before an ant can enter; usually the only way to acquire this is to be reared within the society. In most, but by no means all, ant societies, ants which do not possess a visa are attacked and killed. At first sight the obvious way of acquiring a visa seems to be by descent, so that the visa is genetically acquired. However, since the members of a colony are not genetically identical, even when all are full sisters, there are obvious difficulties about this. At the other extreme the visa might be acquired by living within the society, for instance from the diet of the colony, shared throughout the colony by food transmission. Particularly if different colonies concentrate on different sections of environmental resource, as in honeybees (Kalmus and Ribbands, 1952), this too makes a workable system, though more open to invasion by aliens. The recognition of the visa is a separate problem from its acquisition, since whatever the source of the visa the response to it could be either innate or learned. In fact an ant raised in a colony probably possesses its own innate odour, but learns to recognize a colony visa as a totality to which its own odour contributes a part.

Experiments on the origin of the visa have in general been inconclusive, apparently because the nature and acceptability of the visa does in fact depend on both genetic and experiential factors. In a very full (unpublished) investigation of *Lasius flavus*, Anderson (1970) examined the response of workers to alien colonies. Workers from alien colonies were treated as hostile and subjected to seizing and dragging attacks in 75–80% of encounters. The rate of these attacks fell to about 20–30% in colonies maintained for some time in the laboratory. If a colony was split into several parts which were differently fed (N/8 sucrose, N/8 sucrose + peppermint, honey + black treacle), the rate of attack between members of the parts was only about 5%, though all parts still attacked alien colonies at up to 95%. This seems to show that diet, or at least the diets tested, had little or no effect, and that some genetic basis was involved. Most colony fragments would accept alien queens, but a colony fragment which had been kept with an alien queen was still accepted by other fragments of its original colony. This suggests that the visa originated in the workers rather than in the queen. Mintzer (1982) obtained compatible results from the reverse experiment. He reared colonies of *Pseudomyrmex ferruginea* on genetically identical *Acacia* plants, raised vegetatively in a greenhouse from the same root-stock. Since the ants were entirely dependent on the *Acacia* for their food, the diet of all colonies should have been identical. Separate shoots were colonized with a foundress queen and a group of brood from one of several donor colonies. One month after the brood emerged as workers, the aggression of one group to another was tested. Workers derived from different donor colonies showed a high level of aggression to each other. On the other hand, workers derived from brood of the same donor colony reared by different foster queens did not subsequently behave aggressively to each other. This again suggests that the visa is genetic and resides in the workers and not in the queen. However, the workers are not all genetically identical, so that it is unlikely that they would all produce the same substances. In the analogous case of the stingless bee *Lasioglossum zephyrum*, Greenberg (1979) showed by breeding experiments that the acceptability of workers in alien colonies depended on the average coefficient of relationship (see Chapter 1) of the two colonies. Workers from a colony of cousins were allowed to pass the guards at the colony entrance on about 40% of occasions, while for sister colonies the 'pass rate' was 75–95%. Breeding experiments of this kind would be difficult to set up for most ants, but Carlin and Hölldobler (1983) made up colonies of mixed species of *Camponotus* by cross-fostering pupae into heterospecific nests. They then compared the ratio of hostile (interspecific) to non-hostile (intraspecific) interactions with the ratio expected from the ratio of workers of the two species in the nest. Some species combinations showed significantly too few hostile interactions. Carlin and Hölldobler concluded that the discriminator (in this case at species level) was determined either by the queen or by workers becoming habituated to a collective nest odour, which was in some way the sum of their individual odours, as Greenberg's work suggested for *Lasioglossum*. They argued that if this was so the amount of hostile behaviour should vary continuously with the relatedness of the colonies. If so, the hostility between two workers—sisters from the same source colony, but reared in separate colonies—should depend on the proportion of their sisters in the adopted colony. On the other hand, if visas were not blended in this way, responses should depend only on current colony affiliation. The results (Figure 5.1) agreed almost completely with the second hypothesis. Carlin and Hölldobler conclude that the colony visa

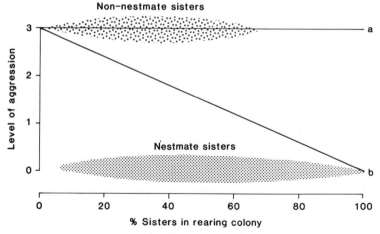

Figure 5.1 The ratio of hostile and non-hostile responses to non-nestmates in colonies with various ratios of two species of *Camponotus*. The horizontal line (*a*) represents the hypothesis that aggression to non-nestmates is independent of their genetic relatedness to the colony; the sloping line (*b*) the hypothesis that aggression falls off with increasing relatedness. (After Carlin and Hölldobler, 1983.)

derives from the queen, in contradiction to the conclusions of Anderson and Mintzer. However, their experiments actually involve interspecific recognition as well as colony affiliation, and the colonies with which they worked had a mean size of 9.3, so that the proportion of sisters used as the independent variable in Figure 5.1 was not very continuous. At present evidence about the origin of the colony visa is therefore rather conflicting: it is quite likely that this has different origins in different groups of eusocial insects.

Murray Blum has recently pointed out the potential importance of cuticular lipids as a medium for nestmate recognition. Their primary function in insects is to waterproof the cuticle, but in some insects, such as the cockroach, low-boiling-point hydrocarbons are continuously secreted into the wax layer and produce the typical smell of this species. Mixtures of similar substances would be well adapted to form the basis of colony odour. In addition, the solid lipid layer itself could act as a medium for the spread and separation of other secretions produced elsewhere.

If the colony visa were to ensure that resources collected by workers are used to increase the inclusive fitness of the workers by promoting the reproduction of close kin, we should expect it to be worker-based. It is a matter of great importance to the workers that their resources are properly directed. It is of course quite immaterial to the queen where her resources come from, or who rears her brood. It is therefore disconcerting to find that most evidence shows unrelated queens to be accepted, as in Mintzer's experiments. Anderson (1970) found that *Lasius flavus* workers often accepted strange queens, though this was more common in queenless colonies and at certain times of year. As in the case of *Pseudomyrmex ferruginea*, fostering by a strange queen did not result in fragments of the same colony becoming hostile to one another. Anderson also found that dead queens were kept or attacked according to their origin, and that this effect was abolished by alcohol extraction and restored by reapplying the extract. In ants, as in honeybees, the treatment of queens by workers depends on a 'queen substance' (see Section 5.5) which also suppresses worker oogenesis and the rearing of replacement queens. This substance is probably responsible for the admission of strange queens into the society, and its existence is evidence of queen manipulation of the worker force (see Chapter 1). However, the adoption of strange queens probably depends on the absence of a more closely related queen, and may demonstrate a strategy of desperation in which worker genes are favoured by the support of any queen rather than none. A strategy by which workers could reinstate the production of brood related to themselves would be by laying unfertilized (male) eggs (see Chapter 1).

Newly emerged workers (callows), whose cuticle is not yet mature, are exempt from the visa system and can be added to alien colonies of their own species. Callows totally lack any aggressive behaviour, and indeed often move very little, resembling and sometimes remaining with, the pupae of the colony. Jaisson (1984) believes they must at this stage lack the specific colony odour, and are themselves unable to recognize it. He suggests that they owe their immunity from attack to their non-provocative behaviour, and also to the production of compounds which inhibit aggressive attacks from the workers attending them. These, or other substances, also attract workers to them, who assist by removing fragments of pupal cuticle or cocoon. This attention by workers of the colony whose aggression is inhibited helps the callows to learn and develop or acquire the colony odour. The cuticle of callows is not fully matured after eclosion from the pupa, and we might expect chemical changes in the epicuticular lipids during maturation. Their lack of a visa conforms with Blum's suggestion that these lipids are important as elements in the visa.

Another set of colony members that are immune from attack as aliens is the brood. Any conspecific brood can usually be fostered. In *Formica polyctena* however, Jaisson (1984) has shown that the ability to recognize even consepecific pupae is not inborn, but has to be learned by the young worker during the first week of adult life. Groups of newly emerged workers were placed with cocoons of another species, and given no contact with cocoons of their own species for 15 days. At the end of this training period, when the young workers had become active and their cuticle had darkened to its final colour, they were allowed to retrieve cocoons of various species from outside the experimental nest. Trained workers first picked up and carried in the cocoons to which they had become familiarized. Later they collected other cocoons, including those of their own species. These were not treated as brood, however, but were eaten. The training was stable, lasting at least six months. This ability to learn at a restricted period of the life-cycle is similar to the phenomenon of 'imprinting' in, for example, young birds. Young workers given no contact with pupae for the first week of adult life were unable to care for brood, whether of their own or another species. The plasticity of the callow worker explains the success of slave-making ants, which kidnap pupae of other, usually related, species. These are allowed to emerge in the nest, and then care for the brood of the slave-maker, to whose brood they have become habituated. The slavers however retain an innate preference for their own cocoons. Errard (1984) has shown that it is easier to set up analogous mixed colonies with congeneric species (e.g. species of *Camponotus*), than with different genera, and that the less related the two species were, the earlier in adult life training had to be begun.

Colony recognition then seems in some cases at least to be based on the genetic constitution of workers. This base is heavily overlaid by acquired behaviour, including habituation and imprinting. If colony recognition has as its function the direction of worker resources into close relatives, why is its basis not exclusively genetic? One reason is perhaps that relatedness within the colony is not uniformly the average of $\frac{3}{4}$ predicted by Hamiltonian theory. Some pairs of colony members are less closely, others more closely, related than this. It is necessary in consequence to establish a global colony visa, perhaps as a cocktail of individual worker's odours, and for individual workers to habituate to this. Secondly, if a colony visa were based on the possession of a certain number of independent genes, a large number would be needed to maintain distinctness among all the colonies in any region. Although on a map only four colours are needed to avoid areas of the same colour meeting along a boundary, this depends on the non-random assignment of different colours to contiguous areas. Such a degree of economy could only be achieved among insect colonies if there were special behaviour to prevent the settlement or persistence of adjacent colonies with similar odour. Apparently not all colonies in an area are in fact different in odour, since even in Mintzer's experiments with the very aggressive *Pseudomyrmex ferruginea* not all colonies fought violently.

Intercolony hostility is reduced in some species such as the wood ants (*Formica rufa* group) and the pest species *Monomorium pharaonis*. In *Formica polyctena* separate nests are connected by trails on which workers pass from one nest to another without challenge, and the complex of nests has been described by Gris and Cherix (1977) as a supercolony extending over an area of 600 ha or more. This assessment is based on the absence of hostility between workers when tested, but there is also evidence that radioactive marked food passes from one nest to another. In *Formica lugubris*, a closely related species, not only workers from nests in the same area, but workers from nests separated by at least 10 km of country without wood ants, are without hostility. Although workers from different nests show little actual fighting, the behaviour of an ant when placed with an alien is not the same as when it is placed with a nestmate. Aliens are subjected to significantly more examination and low-level aggression such as seizing. Whether the absence of fighting is due to the similarities in colony odour, or the presence of a substance inhibiting intraspecific conflict, is not clear. In some other circumstances, usually in spring, both *F. lugubris* (Breen, 1977) and *F. polyctena* (Mabelis, 1979) indulge in 'wars' between colonies.

Mabelis ascribes this to predation, however, and this or territorial conflict, based on special territorial marking, seem more probable explanations than more hostility (Figure 5.2).

5.3 Pheromonal communication

Hölldobler (1978) divided pheromones in social insects, according to the function they served, into *alarm pheromones* which alert colony members to respond appropriately to hazards (Section 5.3.1), *recruitment pheromones* which promote the recovery of resources (Section 5.4), *territorial pheromones* which result in their monopolization, and *sexual pheromones* which bring about the meeting of the sexes or coordinate mating behaviour (see Section 5.5). Although each pheromone may seem to release a particular response, in fact ants and other insects usually practice pheromonal parsimony: in other words, the same substance may elicit different responses in different individuals, or different concentrations, or in different contexts. Generally speaking ants also show glandular parsimony, adapt-

Figure 5.2 Hostility between non-nestmates in *Formica lugubris*. Although overt fights between members of neighbouring nests are rare, workers from more distant nests fight in the laboratory; here two workers attack and kill a stranger from a nest 10 km away. (Photograph by R. Wheeler-Osman, University of Hull.)

ing pre-existing glands to new uses rather than developing completely new glands. The main glands which have been drawn into pheromone production are the mandibular glands in the head and the various glands associated with the sting or the organs derived from it in Formicinae (Figure 5.3).

5.3.1 Alarm pheromones

In many non-social insects, alarm pheromones release dispersal. For instance, when a predator approaches a group of aphids, one of the group releases an alarm pheromone and the group breaks up: as a result, some of its members avoid being eaten or wounded by the predator. Alarm pheromones work in the same way with some ants, particularly those with small weak colonies (Duffield et al., 1976), or which live in tunnels in twigs (*Zacryptocerus*, Olubajo et al., 1980), or in males (e.g. *Odontomachus*, Longhurst et al., 1978). None of these have much to gain from aggressive resistance. However, large colonies with many large workers are more likely to come out fighting in response to alarm pheromones. Very often the glands which produce the alarm pheromones are associated with organs of attack or defence, the mandibles and the sting, and in many cases the pheromones not only serve to signal to nestmates but are themselves repellent or damaging to enemies (Figure 5.3). Compounds like formic acid (which is used as a repellent by the non-social puss moth *Dicranura*) are used as combined alarm pheromone and venom by Formicine ants. Probably the mandibular glands are the most primitive site of pheromone production in ants, and this is still usual in Ponerinae and in many Myrmicinae, though *Crematogaster inflata* is an exception to this, producing an alarm pheromone from its enlarged metapleural glands. Dolichoderinae, *Nothomyrmecia* itself, and *Aneuretus simoni*, the living fossil link between them (Kugler, 1978), employ pygidial (or anal) glands. Formicinae use the mandibular gland and add secretions of the venom and Dufour's gland to it. Release of the pheromone often goes with special postures or behaviour. *Formica* bends its gaster forward between its legs and sprays the products of both glands; *Oecophylla*, *Crematogaster*, and some other Myrmicinae and all the Dolichoderinae lift the gaster forward over the thorax. In all these cases the products act as pheromone and repellent.

Ants employ a variety of classes of compounds as alarm pheromones; a comprehensive list can be found in Blum (1981). The Ponerinae seem to be particularly catholic in their tastes: *Hypoponera opacior* and *Ponera pennsylvanica* use pyrazines, *Neoponera villosa* 4-methyl-3-heptanone,

Gnamptogenys pleurodea 6-methyl-salicylate and the large (and very smelly) *Paltothyreus tarsatus* a number of alkyl sulphides, some of them unique as natural products (Duffield *et al.*, 1976; Crewe and Fletcher, 1974). Myrmicinae tend to employ substances like 3-octanone or 4-methyl-3-heptanone, exemplifying a '3-alkanone theme' (Blum, 1984). Formicinae combine the fairly common use of formic acid with terpenes like citral or their derivatives. Dolichoderinae are characterized by methyl ketones like 6-methyl-5-hepten-2-one. However, there are many differences between species and genera in each subfamily and some genera use unique substances. *Manica rubida*, formerly treated as a member of *Myrmica*, produces 4, 6-dimethyl-4-octen-3-one instead of the 3-octanone typical of *Myrmica*, and this difference confirms its place in a separate genus. Among Dolichoderinae *Tapinoma* uses 2-methyl-4-heptanone and *Hypoclinea* 4-methyl-3-nonanone instead of the usual 6-methyl-5-hepten-2-one. *Azteca* in this subfamily is unique in producing 2-acetyl-3-methylcyclopentene. Some of these substances are also believed to be unique natural products (Blum, 1984). It is not clear what advantages arise from variety in Ponerinae or in Dolichoderinae; unlike sex pheromones, to which only conspecifics should respond, there is no general reason why an alarm pheromone should be species-specific. On the other hand, *Neoponera villosa* and some *Pogonomyrmex* species in the same habitat share the same pheromone, 4-methyl-3-heptanone. Since the stings of both ants are very painful to vertebrates, Duffield and Blum (1973) suggested that these unrelated ants might be part of a mimetic complex. Predators learn to associate a painful sting with the common scent of any of these ants.

5.3.2 *Multiple pheromones*

Many pheromones are used as a blend of several substances, and variation of the mix has probably been the usual way to differentiate, either between different species or, within the same species, to produce variation in response. In *Oecophylla* there are differences in the actual blend of pheromone components between different populations, and even between members of the same population. This is at least in part because secretion of different components is affected by age (Bradshaw *et al.*, 1975). Different components of the pheromone mix have different effects. In *Camponotus pennsylvanicus* the undecane of Dufour's gland is an attractant, but formic acid from the venom gland by itself provokes a non-orientated frenzy. The two together produce a powerful attack on the source (Ayre and Blum, 1971). In *Myrmica rubra* 3-octanone is an attractant and 3-octanol an arrestant (Cammaerts-Tricot, 1973). The two together elicit turning movements and are in addition detected at lower concentrations than when presented separately. This is complicated by the fact that the pheromone is hardly ever applied directly to the responding ant, but reaches it by diffusion or turbulence through the air. When an ant discharges the contents of a pheromone-producing gland, there will at first be a very high concentration of pheromones at the point of discharge. Each of the components will then diffuse outwards along a concentration gradient. We can visualize this gradient as a series of hemispherical shells centred on the ant producing the pheromone. The simplest way these shells could affect the behaviour of a potential responder would be if it responded to concentrations greater than a threshold value. The shell marking this concentration would then enclose a hemisphere of 'active space' in which another ant would respond. Outside the active space it would not respond. Since different substances volatilize and diffuse at different rates, an ant responding to a multiple pheromone will experience changes in the mix with time and also with its distance from the source (Figure 5.4). It is likely to be within the active space for one pheromone component and outside it for another. Of course it is likely that these threshold levels are not the same in all members of the colony, and this might also depend on the state of excitement of the respondent. Response at least in some cases, e.g. *Pogonomyrmex badius*, varies with the concentration of the pheromone: at low concentrations digging is released, at high concentrations attack (Wilson, 1958). The response has therefore to be seen as occurring through a population of ants, with different individuals responding differently according to their position, age, etc. Since a common response is for an ant

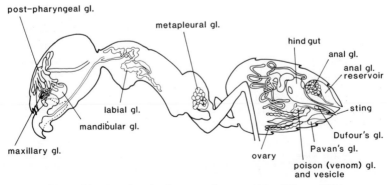

Figure 5.3 Exocrine gland system of an ant. (After Wilson, 1973.)

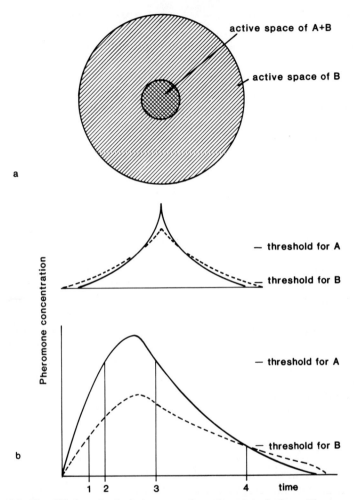

Figure 5.4 The diffusion of a dual pheromone from the point of release. Pheromone A is released in larger amount and diffuses faster than pheromone B. (*a*) Distribution in space; an ant inside the shaded area will respond only to pheromone B, but inside the hatched area it will respond to both. (*b*) Distribution in time; between times 1 and 2 an ant is in active space for pheromone B, between 2 and 3 in active space for both pheromones, and between 3 and 4 only for B. Before 1 and after 4 it is not in active space.

to release more pheromone, the alarm can be propagated through the population. It is therefore very difficult to analyse the effects of a multiple pheromone, or to predict its effects from those of its components.

5.4 Communication in recruitment

Foraging, that is the collection of resources from the environment, has two phases: the search for the resource and its recovery. The search phase is of course the part from which a good deal of the cost of foraging arises; the recovery phase also accounts for some of the cost, for instance in transport, but is the only part which produces a tangible benefit. Clearly there must be advantages in making the ratio of benefits to costs as high as possible, and one way this might be done is by cutting back on search costs. The lowest possible search costs might be obtained by living actually on or in the food source, for instance as a social parasite of other ants (Chapter 7), or as a mutualist on a plant which produces all the colony requires (Chapter 6), though this probably entails other costs—perhaps in defence. At the other extreme, the highest search costs would be produced by sending out a large forager force to search a large area without any information about the distribution of the resources in it. In these circumstances, unless there was a very high density of resource items (seen perhaps as divided up into units just as large as each ant could carry), distributed in a very uniform way, a large proportion of the foragers would fail to find an item and their trip would be a total loss. Any foraging method which reduced the number of ants returning without a load would be an advantage. This argument applies to the case of randomly distributed resource items searched for at random: the argument becomes stronger if the resource tends to occur in patches, since its concentration in some patches must mean that there would be other patches carrying fewer resources or none, from which randomly searching foragers will return without a load. Optimal foraging theory deals with foraging in a 'patchy environment' from the point of view of a single animal which has depleted the resources in one patch and has at some point to search for a new undepleted patch. Ants have the additional option of concentrating their forager distribution, to match the patchiness of resources in the environment, by communication systems which allow the ant that does find a rich patch to capitalize on this discovery by informing its nestmates. The communication could merely inform nestmates that a rich source of food existed, or it might in addition give some indication of its whereabouts. In the second case the rest of the resource patch could be acquired for the discoverer's kin, without further expense in search. There is apparently no reduction in transport costs, which may even be increased by interference between ants and inefficiency in group transport. This sort of response is called *recruitment*. When the discoverer only calls a small group to its find we speak of *group recruitment: mass recruitment* occurs when each recruit itself recruits more ants, so that there is an explosive runaway recruitment of many ants.

5.4.1 *Simple cooperative hunting*

The idea of independent search, which was the starting point of this argument, might seem not to be a real possibility for ants except in a species in which foragers are very thin on the ground, since any predatory insect, searching for prey by detecting movements or by olfactory means, would be affected by the hunting of other members of its species. However, this seems to be in part the case with *Cataglyphis cursor* (Schmid-Hempel, 1984). In this ant, which hunts in arid parts of the Old World, each forager appears to have its own direction of foraging from the nest. It always hunts along this line, using the sun and other visual clues and correcting for the movement of the sun during the day. Individual workers are very efficient at retrieving prey which has a slightly clumped distribution, but there seems to be no way in which they communicate a rich find to other foragers. Other ants which appear to hunt singly, like *Aphaenogaster* (= *Novomessor*) of American deserts, call foraging nestmates over distances up to 1 m by releasing an attractive poison-gland secretion. Another fairly elementary way of using an attractive pheromone for this purpose, found in *Leptothorax acervorum*, is 'tandem calling'. When a forager finds a plentiful food source it returns to the nest and regurgitates food to its nestmates. Then it raises its gaster and extrudes its sting bearing a droplet of a liquid. This attracts nestmates to it in the nest. As soon as the first of these nestmates reaches the caller, the caller runs out of the nest and leads the nestmate to the food source. The follower continually touches the leader with its antennae. If the follower is removed or becomes lost the leader stops and repeats the calling behaviour with raised gaster. Experiments showed that the leader could be kept running by gently touching its hindlegs or gaster with a hair at least twice a second. The follower could be made to follow a dummy treated with poison-gland secretion, but not a dummy with extract of Dufour's gland. A freshly killed ant with the sting gland removed was also ineffective (Möglich, 1975). Hölldobler (1978) believes that tandem calling, and the more elaborate mass recruitment behaviour of other myrmicine ants, has arisen from the application of attractive pheromones produced in the sting gland. (It is interesting that in parasitic myrmicines which have wingless worker-like queens, these attract males by a very similar calling process in which the sting gland secretion is used (Buschinger, 1975): see Chapter 7). Tandem calling has, however, been described not only in other Myrmicinae like *Cardiocondyla venestula* and *C. emeryi* but also in some Ponerinae: (*Ponera eduardi, Bothroponera tesserinodis*) as well as the Formicine *Camponotus sericeus* (Hölldobler, 1978). In the last species, the finder of food lays a trail of hindgut contents during her return to the nest. At the nest she alternates between short fast runs, food-exchange and grooming. Finally, after several brief offers of food the finder returns to the food find followed by some of the ants she has contacted. Usually only one of the latter succeeds in keeping contact with her, and it is usual to see tandem pairs of these ants running in the mid-day sun. The trail of hindgut contents neither guides nor excites followers, it is apparently used only by the scout ant (Hölldobler *et al.*, 1974).

5.4.2 *Group recruitment*

Tandem running usually results only in the recruitment of one companion to the food source. In *Camponotus socius* each ant can recruit 5–30 nestmates. An ant that discovers food places signposts of hindgut secretion round it and returns to the nest laying a trail of the same substance. In the nest it performs a 'waggle display' to nestmates, which alerts them so that they follow it when it returns to the food. This display is essential for recruitment, and scouts whose glands have been closed with wax are still able to evoke the following response. Outside the nest these recruits are able to follow the hindgut trail. In the absence of a leader, however, they only persist along it for 10 m and do not usually find the food (Hölldobler, 1971). In *Formica fusca* the course of recruitment is similar except that recruits will follow the pheromone trail in the absence of a leader (Möglich and Hölldobler, 1975) and *Camponotus pennsylvanicus*, though normally recruited by a motor display or dance, will follow a hindgut pheromone trail without this preliminary (Traniello, 1977). Tandem running shades in these species into a rather short-lived "group recruitment", using a trail pheromone as its means of communication.

5.4.3 *Mass recruitment*

Trail recruitment can, however, be much more imposing than tandem running. Many ants, such as *Solenopsis invicta* and *Monomorium pharaonis*, use less fugitive trails to recover large and unpredictable food patches. A relatively fixed trail is formed, which carries hundreds of workers to and from the food, and persists until the food is exhausted. It has been known since the early part of this century that this trail is a chemical track, since ants follow it when it is displaced for instance by turning a piece of paper over which ants are running. More recently, Wilson (1962) was able to show that in *S. invicta* the trail consisted of smears of the secretion of Dufour's gland made by dragging the tip of the sting along the ground. The pattern of diffusion of each pheromone smear is similar to that of an alarm pheromone. However, in this case there is a line, usually discontinuous, of spots all laid down at different times and diffused outwards to different extents (Bosset and Wilson, 1963). The 'active space' in this case is semi-ellipsoidal, and probably an ant which is inside the active space responds by running. If its movement takes it out of the active space, however, it turns so as to re-enter the space (Figure 5.5). *Lasius fuliginosus* workers can follow trails after one antenna has been amputated, but make errors towards the operated side. Workers with antennae glued in a crossed position also follow, but with even more difficulty. This suggests that they can use several mechanisms of orientation within the active trail space (Hangartner, 1967). If the trail is followed in this way it is hard to see how the ant can detect whether it is running 'up' or 'down' the trail. There is some evidence that ants on a trail are also orientating to stimuli like light or gravity, which can supply this missing cue. In *Acanthomyops subterraneus,* Hangartner (1969) showed that the richness of the food source affected the amount of trail pheromone laid. This shows that trail-laying can be modulated to carry other information, though Hangartner did not show that other ants responded to this.

Trail pheromones have been more difficult to identify chemically than alarm pheromones. Some of them at least are more complex than many alarm pheromones: *M. pharaonis* uses a mixture of 5-methyl-3-butyloctahydroindolizine and other compounds, and *Iridomyrmex humilis* (Z)-9-hexadecenal and other substances (Haynes and Birch, 1984). The physical characteristics required in a trail pheromone are rather different, since it is applied to the ground and is often more persistent. In real life each ant that returns along the trail having fed adds new pheromone to the trail, so that the trail persists until the food is exhausted. In *S. invicta* the short persistence of trails on a glass substrate in the laboratory suggested that a single worker would not be able to guide recruits more than 1 m without trail reinforcement of this kind. Since each successful ant which returns to the nest recruits more ants to the trail, the number of ants involved increases exponentially at first to level off asymptotically. The levelling could be either because all the available ants have been recruited, or because the rate at which ants can get access to the food limits the rate at which fed ants return. In either case the numbers would follow a logistic curve (Figure 5.6). Very often the number of visits to the food shows a slight overshoot, i.e. the last ants arrive when the food is already exhausted. The latecomers usually search around in the neighbourhood, and if food is somewhat clumped in its distribution they may find again and start a new recruitment process. Some inaccuracy in the guidance of ants to the food may similarly be seen as adaptive.

In recruitment, as in the case of tandem following, mechanical signals as well as purely chemical ones seem to play a part. Although Wilson (1962)

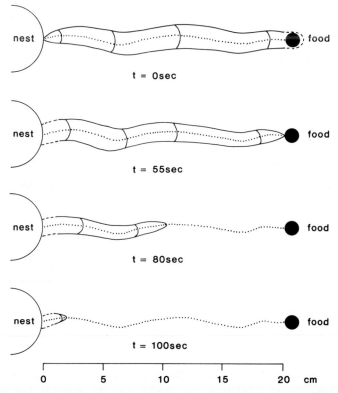

Figure 5.5 The development and decay of the active space of a trail pheromone. In this case the trail persists just long enough for the ant laying it to reach the nest. (After Wilson and Bossert, 1963.)

showed that *S. invicta* workers could be made to leave an artificial nest *en masse* by blowing Dufour's gland extracts (the trail pheromone in this species) into the nest, in many other species (*M. pharaonis*, Sudd, 1957; *Pheidole* sp., Szlep, 1970), when successful forager ants return to the nest they run rapidly and strike other ants with their legs or antennae. It is difficult to show the function of these movements by experiment, but they may recruit workers (as in the group recruitment of *Camponotus socius*), or merely help physically in spreading a pheromone through the tunnels of the nest. Actual contact with recruits might allow the recruiter to monitor the response to its chemical signal, and could be a prerequisite for its own return to the food.

5.5 Sex pheromones

Just as alarm or recruitment pheromones are produced primarily by workers and act to coordinate the behaviour primarily of workers with each other, so sex pheromones are produced by reproductive male or female ants and act to coordinate their behaviour with each other and with that of workers. The main problems solved in this way are (i) the suppression or control of worker reproduction; (ii) the coordination of the mating flight and (iii) the discovery of one sex by the other.

5.5.1 *Queen pheromones*

As in the honeybee, queen ants are able to suppress the production of other sexual forms by the workers. In *Solenopsis invicta* a non-volatile substance produced from the mandibular glands of the mated queen prevents ovarian development and wing-casting in young queens, and also inhibits workers from rearing larvae as queens rather than as workers (Fletcher and Blum, 1981). A second substance produced by the queen is volatile, and attracts workers to the queen. In *Myrmica rubra* a similar secretion interacts with secretions from potential gyne larvae at a critical age, and controls the way the larvae are treated by workers. The larva produces a substance from a glandular area on its ventral surface; in the absence of a queen such larvae are fed freely and continue to grow, eventually becoming fully reproductive winged females. In the presence of one or more queens, however, this substance provokes the workers to attack such larvae, leaving characteristic scars. The attacks stunt the further growth of the larvae at this apparently critical stage, and they become workers. Attacks are also provoked by administering queen head extracts (Brian, 1970).

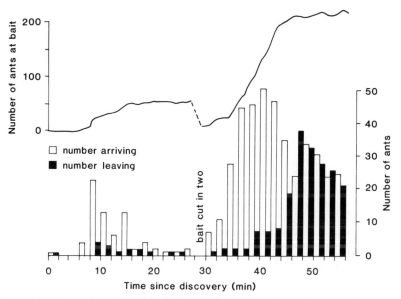

Figure 5.6 The number of ants at the food source in mass recruitment. The curve is logistic in form; its rate of increase depends on the return of trail-laying scouts. (After Sudd, 1963.)

5.5.2 Other sexual pheromones

With the exception of army ants and some parasitic ant species, the males and females mate outside the colony. In many species quite large numbers of males and/or females are produced, either as a part of the normal development of the colony over a period of years, or as a response to the death or inadequacy of the existing queen. The details of the mating strategies used by ants have not really been worked out. Biologists have perhaps been fascinated in the past by the importance of outbreeding in maintaining genetic diversity, and mating outside the colony obviously makes this possible. The fact that some species (for instance *Atta capiguara*, Amante, 1972; *Lasius niger* and *L. flavus*) mate after flights in which colonies over a large area of country take part on the same day or at the same hour, looks like an arrangement to promote outbreeding. Other species, however, mate after quite a limited flight in which queens must be more likely to mate with males from their own nest or from a nest close to it. Since the haploid male will contribute all his genes to a daughter, mating with a sister ($\frac{1}{4}$ relatedness) produces a relationship of $\frac{5}{8}$ with his daughters. Particularly in species where males are produced from eggs laid by workers, it could be in the interest of workers too to promote mating between their sisters ($\frac{3}{4}$ related) and brothers ($\frac{1}{4}$ related) to produce founder queens ($\frac{1}{2}$ related), rather than matings with unrelated males. Presumably this is balanced in some way against the possible disadvantages of inbreeding. In *Camponotus herculeanus*, males begin to sun themselves at the nest entrance as much as two months before swarming. They are later joined by females, and eventually both sexes leave the nest in an avalanche (Hölldobler and Maschwitz, 1965): this may affect inbreeding. This synchronization is due to the release by males of their mandibular gland secretion, which contains mellein and methyl 6-methyl-salicylate. This pheromone increases the activity of females markedly, so that they run up the tree to the highest twigs of the crown. It has only a slight effect on workers and no effect at all on males (Payne *et al.*, 1975). Since the workers probably delay the departure of the females, the effect is at first to prevent females leaving and then for them to leave in a sudden crowd. In *Pogonomyrmex* on the other hand the males leave the nest first and assemble at specific sites, the same ones from year to year. In *P. desertorum* and *P. maricopa* the males assemble on desert bushes and in *P. rugosua* and *P. barbatus* on the ground. At these sites males release 4-methyl-3-heptanone (or its alcohol) from their mandibular glands. This attracts more males and also females by upwind flight. The species are separated by the time of flight and by using different assembly sites. When the females arrive at the assembly site the males are attracted to them by a species-specific pheromone from the (female) poison gland and by a surface pheromone (Hölldobler, 1976*b*).

Where there is no flight the males' pheromones seem to be absent or unimportant, and the female attracts the male in or near the nest. In *Monomorium pharaonis* the male is attracted from up to 60 mm by the Dufour's gland secretion of the female. When he approaches within 10–20 mm the same secretion excites him and he will attempt to mate with anything of about the right size, though successful mating also needs some other, probably tactile, signals from the female. *M. pharonis* is most common as a pest in heated buildings and mates inside its nest, but originally it was probably a parasite (see Chapter 7). In other parasitic ants like *Leptothorax kutteri*, *Harpagoxenus sublaevis* and the commensal *Formicoxenus nitidulus* (Buschinger 1975, 1976), the female calls the male in a similar way, often standing with her gaster extended and her sting extruded for some days. Species in which the female is ergatomorphic (i.e. never develops wings and resembles a worker) also have to attract the winged male to the female. In the ponerine *Rhytidoponera metallica* the female exposes a tergal gland between the last two gastral sclerites. Males fly upwind at low concentrations of its secretion. In higher concentrations they become excited and will attempt to mate with each other as well as with females (Hölldobler and Haskins, 1977). In *Megaponera foetans* the winged male searches for and enters other nests, apparently following old foraging trails (Longhurst and Howse, 1979), and similar behaviour is known in Dorylinae (Schneirla, 1971).

CHAPTER SIX

ANTS AS PARTNERS

A very simple-minded approach to ant biology might suggest that an ant colony is a society at peace within itself, but at war with the rest of the living world. As we have already seen in Chapter 1 the first proposition is not strictly true, since even within the cooperative world of the colony there are tensions and conflicts. It should come as no surprise, then, that ant societies are not simply pillagers and exploiters but are in mutualistic relations with many other organisms. Mutualism is a term used to describe a relationship between two species which results in a net benefit to both. Strictly, the benefit should be an increase in genetic fitness; however, it is often almost impossible to show that this is the case and we have to adopt the relaxed requirement that the relationship increases the rate of population growth of both. Seen in this way, mutualism fits into a set of relationships which also includes predation and competition. In competitive relations the two species interact to reduce each other's population growth, and to this extent competition is immediately detrimental to both. On a longer time-scale, competition may increase the fitness of one or both species by selection. Predation has in the short view a benefit to the predator which can increase faster when there is plenty of prey. The rate of increase of the prey, on the other hand, is reduced when predators are plentiful, so that predation is asymmetric with benefits to one partner and detriment to the other. On a longer time-scale predation controls the population of the prey so that it does not over-exploit its resources, whilst over-plentiful prey may lead to growth of predator populations and to an eventual reduction in prey numbers. Predation thus acts to control the sizes of both populations, giving stability to the system. In mutualism each species is able to increase faster than it would in the absence of the other. A system of this kind shows positive feedback, since an increase in one mutualist partner should benefit the other and cause it too to increase. The relationship should be unstable, and lead to over-large populations of both.

6.1 Ants in the ecological community

Another elementary principle of ecology is that organisms become less abundant the higher their trophic level (the pyramids of numbers and of biomass). In general herbivores have a smaller biomass than the plants on which they feed, and predators in turn a smaller biomass than their prey. Since ants are in general carnivores they should therefore not be too abundant, and this conflicts with the idea that the benefits of social life might increase with the size of the society. In fact we find that at least in some habitats ants are very numerous and may make up a large proportion of the total biomass. In West Africa Lamotte (1947) found that ants constituted 27% of invertebrate numbers and 4% of biomass. In North American grasslands foraging ants contribute 1–15% of the total biomass of above-ground invertebrates, and 17% of the numbers of subterranean invertebrates. The proportion of ants in European grasslands is also high, 1–2% of invertebrate biomass, and is large in relation to that of other invertebrates such as spiders (Pisarski, 1978).

The arithmetic of the pyramid of biomass conforms to the Second Law of Thermodynamics, which states that transformations of energy can never be 100% efficient, and always result in the production of heat. This applies to the transformation of plants into herbivores, for example; only a small proportion of the energy contained in plants can be converted to herbivores, and little of this in turn to carnivores. The energy content, and as a consequence the biomass and numbers, decrease at each stage. Carnivorous animals are at least two stages along this line, and consequently cannot exploit more than a small fraction of the energy which enters their ecosystem. The size of this fraction can be increased only by moving nearer to the plants or primary producers, and exploiting them more directly, after fewer transformations of energy. Ants, though fundamentally carnivorous, often function as both primary and secondary consumers (herbivores and carnivores) (Pisarski, 1978) by entering into specialized and intimate mutualistic relations directly with plants (as in myrmecochory and in myrmecophytic plants) or with herbivores. Usually the herbivore is a homopteran, but sometimes Lepidoptera are involved.

6.2 Ants and plants

Probably the relation of bees to the flowering plants they pollinate is the best-known example of a mutualism. The evolution of bees has been closely paralleled by the evolution of floral structures which advertise the rewards available to pollinating insects. In mutualisms with ants, some plants confer the benefits of food on their ant partners by means of extrafloral nectaries. Other plants, which supply a nesting site, are called myrmecophytes; they often offer food as well. The ants in turn confer on their plant partners the benefits of protection from predators or competitors, transport of seeds (myrmecochory) and sometimes nutritional benefits as well (Beattie, 1985).

On the whole, ants are not heavily involved in pollination of flowering plants: ant pollination of some plants is occasionally reported, but the lack of flight makes ants poor, though possibly economical, as agents of outbreeding. The self-sterile herb *Polygonum cascadense* seems to be specialized for ant pollination in 10 ways (Hickman, 1974). It lives in a hot, dry habitat in dense stands poor in species; it is short or prostrate with accessible nectaries on sessile flowers of which not many are open at the same time. Its pollen is very sticky and small in amount, to reduce the chance that ants will groom themselves on the source plant. The amount of nectar is also small, so that ants visit up to 20 flowers before returning to the nest. This also discourages flying insects from visiting. The system is probably cheap from the plant's point of view, but the requirement that flower opening is not synchronous on a single plant must delay the rate of seed set. Some alpine plants also depend on ant pollination, perhaps because wind-speeds limit the flight of bees. Pollen on less specialized plants can probably be dispersed just as well by simple shedding as it would be carried by ants. Many flowers are indeed specialized to exclude ants; thefts of nectar by ants from *Asclepias curassavica* reduce the duration of visits to the flowers by butterflies and the proportion of butterflies to which pollinia are attached (Wyatt, 1980).

6.2.1 *Ants and extrafloral nectaries*

The fact that many plants possess extrafloral nectaries is not well known. The nectaries are glands on the leaves or stem, and are most frequently visited by ants. The ants obtain a concentrated and immediately assimilable energy source, and protect the plant from herbivores. The plant does not provide the ants with a protein source in its nectar, and this compels them to prey on insect predators of the plant. The balsa tree *Ochroma pyramidalis* is pollinated by bats, and produces a floral nectar with a sugar content equivalent to 11% sucrose, and rich in amino acids, to attract them. Its extrafloral nectaries, on the other hand, contain a more concentrated nectar with little amino acid. The extrafloral nectaries are visited by ants; if the ants are excluded experimentally there is an increase in the amount of insect damage to the leaves. The energy content of the nectar is about 1510 J per leaf, and about 13 cm^2 of leaf could be produced for this expenditure of energy. This is equivalent to about 1% of the area of a leaf, so that if more than 1% of leaf loss is prevented by ants the nectaries are cost-effective (O'Dowd, 1979). The secretion of the extrafloral nectar is often scheduled at the time and site of greatest risk of damage. In *Bixa orellana* the extrafloral nectaries secrete only as the bud is opening and is most susceptible to damage (Bentley, 1977). In *Costus woodsoni* the nectaries are placed and timed to protect seeds from the fly *Euxesta* (Schemske, 1980). Bracken-fern *Pteridium aquilinum* has nectaries on its fronds which are visited by wood ants; young fronds are protected by cyanogenic glucosides, old ones by tannins, and the gap between the two is closed by the secretion of a dilute nectar which is attractive to wood ants early in the year (Lawton and Heads, 1984). As we shall see, plants also receive similar protection from ants which are tending aphids and other Homoptera (see Section 6.5).

6.2.2 Ants as allelopathic agents of trees

Predaceous ants can give any plant some protection from predators if they hunt on its surface. Extrafloral nectaries attract ants to susceptible sites at susceptible times. Some trees in the tropics take this relationship a good deal further and use ants as a defence from competitors as well as from predators. The bulls-horn acacias (*Acacia dadada*) of Central America have enlarged hollow thorns. These are usually inhabited by *Pseudomyrmex* colonies, founded by a queen who gnawed her way into the thorn. The *Acacia* also provides food from nectaries and from Beltian bodies on the leaf-tips. The latter are rich in protein and lipids, and the *Pseudomyrmex* do not depend on predation for a protein source. They are, however, extremely aggressive and repel insect and mammalian attacks on the tree. More important, they attack any other plant which touches their tree, and any plants which grow within a circle below its canopy. The *Acacia* is typically a tree of secondary bush, and the removal of competitors by the ants allows it to grow rapidly and secure a place in the canopy (Janzen, 1966). The relation between another prominent Central American plant (*Cecropia*) and the *Azteca* that live in its hollow stems is more puzzling. *Cecropia* is also a tree of secondary regeneration; it has hollow stems which *Azteca* enters through weak points at each node. Below each bud is a pad-like Müller's organ which provides food too. Although the ants are unpleasant to human attackers, they seem to give little protection to the tree from insect pests, and undamaged trees without ants and damaged trees with ants are quite common. Janzen (1968) showed that the ants attack the growing tips of vines which touch the tree and also remove the seedlings of epiphytes, which can only attack *Cecropia* above the altitudinal range of *Azteca*. Janzen compares the benefits of ant occupation to the possession of allelopathic chemicals, which prevent the growth of competitor plants.

6.2.3 Transport of seeds by ants

Seeds are a much more concentrated food-source than the vegetative parts of plants, and many of the problems of the digestion (or non-digestion) of cellulose are avoided by granivorous animals. Ants in desert and steppe often depend on seeds for a major part of their diet (see section 2.2.5). These harvester ants, however, eat the seeds that they collect: they may aid plants by thinning concentrated patches of seed which might be very susceptible to predation by rodents. In general, however, the removal of seeds from the 'seed-bank' of desert plants, where seeds may wait up to seven years before germinating in unpredictable rainfall, is detrimental to the plants. More specialized plants in a number of other habitats, from tropical rainforest to alpine tundra, have seeds which are attractive to ants. There is a lot of evidence that the lipid content of the seed is important in attracting ants, and most ant-borne seeds have a specialized oily organ, the elaiosome. Ants collect the seed, but eat only this oily organ, often an aril, leaving the seed embryo and its other reserves intact. The seed is transported towards the nest, but discarded after the ants have eaten the elaiosome.

The seeds of some species of *Viola* are partly distributed by ants. Several American species retain the ability to eject their seeds when the capsule bursts, but others, and apparently most Eurasian species, have lost this power. The amount of ant dispersal of the seeds in woods in West Virginia depended very much on the ant fauna present: the most effective ants were *Aphaenogaster* spp., especially *A. rudis* and its close relations. By removing the seeds from the neighbourhood of the plant, ants help to reduce the effect of very heavy rodent and bird predation. The actual distance the seed is moved is not impressive, but by discarding seeds near to the nest, in soil freshly brought up from below the humus layers, ants improve germination in a safe site (Culver and Beattie, 1978). *Sanguinaria canadensis*, in a similar habitat, has no other mechanism of seed dispersal, and is to that extent an obligate myrmecochore. Again the efficiency of ant transport varies from place to place, being much less where *Aphaenogaster rudis* is absent. The seed is fairly large and has a large white elaiosome; the size of *A. rudis* seems to be an important factor in its effectiveness in dispersing seeds, and in experiments it collected over 70% of seeds. Where it is present, *Sanguinaria* is more evenly distributed; each clump has more stems and produces fewer seeds per seed-capsule (Pudlo *et al.*, 1980).

6.2.4 *Ant gardens*

The epiphytic vine *Codonanthe crassifolia* (family Gesneriaceae) has a different relation with ants like *Crematogaster longispinosa* in Central American lowland forest. The vine roots in debris on the stems of trees, especially the pejiaye palm *Bactris utilis* grown in plantations for its fruit. Although *Codonanthe* has floral and extrafloral nectaries, it is probably not pollinated by ants and gets little protection from damage by flea-beetles. *Crematogaster longispinosa* collects the vine fruits and eats the pulp. The seeds are built into the walls of its carton nest, where they germinate. In experiments seeds would not germinate on the ground and would not grow on damp cotton or on palm bark; they need detritus or the ant carton. Since the ants feed on insects, honeydew and nectar, nutrients are brought into the system from outside. The advantage of the mutualism to the ants is not very clear, but *Codonanthe* nectar seems to be an ingredient of the carton and the roots of the epiphyte may also strengthen the nest (Kleinfeldt, 1978).

6.2.5 *Ants and epiphytes*

In the tropical forests of the Far East some epiphytic plants have a very different relationship with ants, in which, far from the ants feeding on the plants, they carry in food from elsewhere. The residues of this provide the epiphyte with nutrients. About 40 species of *Hydnophytum* and 25 species of *Myrmecodia* (family Rubiaceae) are obligate nutritive myrmecophytes in this way (Figure 6.1). As the seedling epiphyte begins to grow, a tuber develops on the epicotyl; this later develops a system of tunnels with pigmented walls and a warty lining. Ants nest in the tunnels, getting in and out through rings of very small pores on the outside. The presence of ants increases the growth of the epiphyte, and radioactive marked nutrients placed in the tunnel are absorbed by the plant. The ants are not aggressive species and apparently afford no protection from insect pests, since they usually feed as scavengers. In tropical America, Bromeliads have a similar relationship with ants; in the myrmecophytic species the leaves are closed to prevent water getting in (Huxley, 1980). These ants and their plants are a very striking example of mutualism allowing the exploitation of a difficult habitat. In the forest canopy the plants' problem is having access to light without growing down to the soil for nutrients; the ants' problem is having

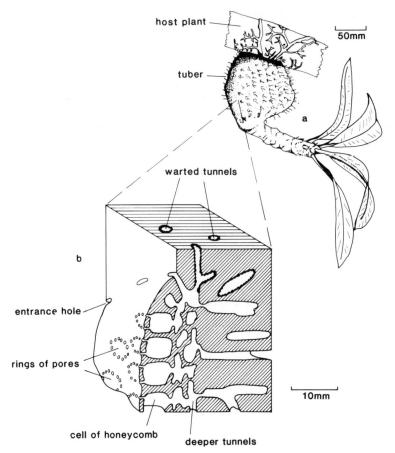

Figure 6.1 The ant-epiphyte *Myrmecodia tuberosa*. (*a*) Whole young plant; (*b*) schematic block view of tuber: (i) 'cell' of 'honeycomb', (ii) ring of pores, (iii) entrance hole, (iv) warted tunnel, (v) and (vi) deeper tunnels. (After Huxley, 1980.)

access to the actively photosynthetic parts of plants without returning to the soil to nest. The accommodation between the two solves this neatly.

6.3 Ants and other insects

Ants, basically carnivorous, are only able to feed on plants where there are special features of the plants (nectaries, Müller's organs, Beltian bodies), which supply the ants with usable food. These features apparently arose by co-evolution in a mutualistic set-up which benefits both ant and plant partner. Ants' mutualisms with other insects are strikingly different, since the commonest partners (aphids and other Homoptera, caterpillars) are just the sorts of insects on which ants feed. They are soft-bodied, relatively defenceless, often aggregated in groups and engaged in converting plant material to nutritious animal protein. Their relationship with ants is therefore concerned as much with 'turning off' ant predation as with 'turning on' the useful services which ants supply. In general the ants obtain food, especially carbohydrates, from their partners in return for some hygienic services, for protecting them from enemies and, in some cases, providing a nest.

The food which ants get from aphids and scale insects is honeydew. Unlike the secretions of the nectaries of plants, which are specially produced to attract mutualists, honeydew is a by-product of the feeding method of Homoptera. Homoptera have suctorial mouthparts (stylets) with which they can pierce the cuticle and cell-walls of plants and extract cell contents. Aphids are able to control the movements of their stylets within the plant, and in many cases can place them in particular cells. Phloem vessels are often attacked, and this has the advantage that the aphid can tap the main transport stream of the plant. In addition the contents of the phloem are under pressure, whereas other vessels of the plant, such as the xylem, are under tension. The alimentary canal of many Homoptera is specially adapted to cope with this liquid diet, in particular to eliminate the large amount of water it contains. The excess water and much carbohydrate is eliminated as liquid faeces, known as honeydew. As well as water and sugars this contains quite large quantities of amino acids and other substances. In summer aphids increase their numbers by viviparous parthenogenesis, and can form very dense clumps. These are susceptible to predation as well as to the effects of pollution by honeydew, which encourages fungal disease. Ant attendance brings them the benefits of protection from predation and the removal of honeydew.

6.3.1 *Ants and aphids*

Honeydew is the food of a community of various organisms, including bees, various flies and wasps, and moulds, as well as ants. It also serves as a clue to some aphid predators, such as syrphid flies, which are attracted to it to lay their eggs where their young can prey on aphids. Ants differ from all the other members of the community because only they collect honeydew directly from the anus of the aphis, instead of licking fallen honeydew from the plant surface. This has often been called 'milking' of aphids, but in reality it is more complicated. Each aphid produces a small droplet of honeydew at intervals of a few minutes. There are usually fewer ants than aphids, and each ant services several aphids, passing from one to another. In *Symydobius oblongus* tended by *Formica lugubris* on birch, Douglas and Sudd (1980) found that each aphis had an ant in close attendance for 14% of the time, but that it produced 84% of its droplets while the ant was in attendance. By studying the time relations between the movements of the aphis and the ant, they were able to show that the aphis was more likely to defaecate soon after an ant touched its abdomen, which fits in with the description of ants 'milking' aphids. They also found that patrolling ants responded when the aphis raised its abdomen, a preliminary to defaecation, so that the aphid was able to call an ant to service it. These two communication systems help to account for the efficiency of honeydew collection; 84% of droplets were collected by the attendant ants, and the figure is probably higher than this for most of the year. In *Formica lugubris* honeydew makes up about 75% of loads carried back to the nest by ants, and it is clear that ants receive a benefit from this.

Aphid attendance is rather more common in the ant subfamilies Dolichoderinae and Formicinae, which have better developed crops and proventricular valves. Ants in other subfamilies do, however, attend aphids, and even some Ponerinae carry nectar or honeydew in their jaws. In most cases that have been studied an ant species is not tied to a particular aphis species. In Europe 22 species of subterranean aphids are cultured by 17 species of ant. *Lasius flavus* was associated with 13–17 species of aphids, but less specialized ants might only have 1–4. In West Scotland, however, Muir (1959) found that many nests contained a single aphis species. The advantage of the association to ants is clearly nutritive. Ants use aphids as a device for tapping the phloem stream of plants. Since the plant materials. have not undergone much chemical transformation by the aphid, little energy is lost by aphid respiration. In effect, therefore, ants are living as primary consumers and increasing the amount of energy available to them from the ecosystem. In Bavaria, Horstmann (1974) has calculated that wood ants obtain $250\,\text{kJ}\,\text{m}^{-2}$ from honeydew, but only $96\,\text{kJ}\,\text{m}^{-2}$ from prey. This much-increased energy input may make it possible for ants to hunt prey in more extravagant ways (for instance over greater distances), and to use energy in constructing nests, etc. (see Chapter 3). It is clear that many ants also eat the aphids they are tending. A species of ant like *Lasius flavus* has a very high population density; as it feeds almost entirely underground it must depend on root-feeding insects for much of the protein in its diet. Pontin (1961) found that in midsummer most of the food this species fed to its larvae was aphids it normally tended, but that no aphids were found in the prey before June. A later study of the populations of these aphids on grass roots showed a large deficit in the number of first-stage aphids, compared with what was to be expected from the numbers of adult parthenogenetic females. The conclusion that ants are culling young aphids is inescapable (Pontin, 1978). Way (1963) came to a similar conclusion from his studies of scale-insects tended by *Oecophylla longinoda*; as the population of scale-insects increased, the ants began to eat them. It is not clear how the ants balance the benefit they obtain from consuming some of the aphids against the benefits of collecting honeydew, and we will return to this point later.

Some aphids like *Aphis fabae* can benefit from ant attendance, but survive equally well without it. These are facultative myrmecophiles. Some others like *Drepanosiphum pseudoplatanoidis* are never attended and are only preyed on by ants. Many other species are never, or rarely, found without ant attendance; these are obligate myrmecophiles. Few of them are tied to a single species of ant; for instance *Pterochloroides persicae* in Israel is attended by 11 species of ant, belonging to 10 genera and 3 subfamilies. Subterranean species like *Paraclitus cimiciformis* are probably limited to ants like *Lasius flavus* which specialize in underground tending (Sudd, 1987). Specificity may be higher in tropical aphids and other Homoptera about whose relation with ants we still know very little. Aphids which are only facultatively or never attended by aphids have long siphuncles which discharge a waxy defensive secretion and alarm pheromones. They may also have their anus placed on a cauda which flicks away the faeces (Figure 6.2a). Obligate myrmecophilous aphids typically have reduced siphuncles (Figure 6.2b), and a perianal ring of bristles which holds honeydew on the aphid until the ant removes it. They therefore show anatomical and behavioural adaptations to being tended by ants: they use their repugnatorial glands less, and do not actively cast away honeydew.

Facultatively myrmecophilous aphids usually have larger populations when they are attended by ants and often have larger colonies (Way, 1963). Addicott (1978a) made a detailed study of the fate of attended and unattended aphis colonies of four species on *Epilobium* in Colorado. The aphids were attended by 10 species of ant. He found that the probability that an aphid colony would survive on a marked *Epilobium* shoot for two weeks was significantly altered by the presence of ants. Colonies of *Aphis varians* and *A. helianthi* were more likely to survive, but colonies of *Macrosiphum valerianae* less likely, if attended by ants. The effect of ants

Figure 6.2 Anatomical differences between aphids attended and not attended by ants. (*a*) A non-attended aphid on *Aquilegia* has long legs, elongated siphuncles with a flared rim and its anus is on a process, the cauda; (*b*) the obligate myrmecophile *Symydobius oblongus* on birch has shorter legs and its siphuncles are low ring-like warts. Scale bar 200 μm. (Scanning electron micrograph by C. Ellis, University of Hull.)

also depended to some extent on the size of the aphid colony: generally ants had a more marked beneficial effect on small colonies than on large ones. This is an interesting finding, since unbridled mutualism in theory tends to cause indefinite increase in the numbers of both partners. If the benefits of mutualism were density dependent, however, the mutualistic system would be more stable.

Obligate myrmecophiles, by definition, do not occur without their ant mutualists, so it is not possible to compare the performance of attended and unattended colonies. Aphids like *Cinara pini* and *C. kochiana* seem to be dependent on attendance by wood ants. *C. piceae* and *C. bogdanowi* have been found both attended and unattended, and seemed not to differ in population size or phenology (Sudd, 1987). When *Formica polyctena* was introduced from Italy into Quebec, after two years it was found to be attending 21 species of aphis, seven of them not previously found in Quebec (McNeil *et al.*, 1977). Whether the new species were previously present in low numbers or were new immigrants is not known, but clearly the introduction of a suitable ant species allowed them to form large populations. Subterranean aphids, like those tended by *Lasius flavus*, are a special case. Unlike the species of *Cinara* tended by *Formica lugubris*, which have only a single woody host plant, aphids like *Anoecia* and *Tetraneura* have a basically dioecious cycle, spending summer on the roots of grasses or herbs and returning to a woody host in winter. *Anoecia corni*, which is attended by ants, retains this habit, and is said to be holocyclic. It is not found underground in winter, having returned to its primary host *Cornus* to lay overwintering eggs. Other species like *Baizongia pistaciae* and *Tetraneura ulmi* are found below ground in winter, and have become anholocyclic, not returning to their respective hosts, *Pistacia* and *Ulmus* (Pontin, 1978). Anholocyclic aphids like these seem to be totally dependent on their ants for surviving winter, indeed *B. pistaciae* survives in Britain where its primary winter host is absent. *Lasius flavus* carries aphid eggs into its nest in winter (Pontin, 1960a), but the evidence that these eggs are replaced on the host plant in spring is not yet forthcoming.

6.3.2 Ants and Lepidoptera

The second group of insects with which ants have mutualistic relations is the Lepidoptera, especially the family Lycaenidae. This family contains 40% of all butterfly species, and its success and diversity may be to some extent at least connected with its relations with ants since 245 of its 833 well-described species have some relation with ants. The caterpillars are short-legged and compact, and many species have hairy glandular cuticle and a gland in the middle of the back which attracts ants. Unlike Homoptera, which have a waste product of potential value as a result of feeding on plant sap, most caterpillars eat the cells of plants. They are unable to digest the cellulose cell-walls, which are passed as very voluminous dry faeces ('frass'). Lycaenidae, however, have specialized in feeding on the nitrogen-rich parts of plants, such as buds and fruits (Pierce, 1984). Some species have abandoned phytophagy for part or all of their larval life, and feed on Homoptera or on ants (Cottrell, 1984). Like aphids, caterpillars are much sought as prey by insects and other predators; many have repugnatorial glands and other chemical weapons. These weapons are less effective against parasitoid insects, and it is from these that they most need defence. Pierce and Mead (1981) showed that this was the case in the Lycaenid *Glaucopsyche lygdamus* feeding on *Lupinus* in Colorado, where it was attended by *Formica altipetens*. Tending began at the third instar, after which all larvae were attended. Each caterpillar had an epidermal gland on the abdomen which was licked by ants. Caterpillars from which ants were excluded experimentally had significantly higher rates of parasitization than those with ants. Ants afforded much more protection against braconid wasps than against tachinid flies. Caterpillars also had an osmaterium, producing repellant substances. In *Lysandra hispanica*, the Provence Chalk-Hill Blue butterfly, larger larvae tended by *Plagiolepis pygmaea* can produce fluid from their gland about every two minutes for at least an hour. The fluid contains 10–20% total carbohydrate, mostly fructose, saccharose, glucose, trehalose and three other higher sugars. Trehalose is absent in the sap of its food plant *Hippocrepis*, and the total carbohydrate in the caterpillar's blood is only about 2%. It is clear therefore that the caterpillar is investing energy in producing the fluid (Maschwitz *et al.*, 1975). Pierce (1984) showed that the secretion contained high concentrations of sugars and of the amino acid serine. A synthetic soup of this amino acid was very attractive to *Iridomyrmex* spp. that tended a Lycaenid, but not to *Pheidole megacephala* which did not. Lycaenid feeding habits involve some concentration of caterpillars on specialized parts of plants, and this may expose them to enemies (Atsatt, 1981*a*).

The mother butterfly is responsible for placing her eggs near ants. *Ogyris amaryllis* females lay preferentially on mistletoes that have ants on them, even if those plants are nutritionally inferior species or totally unsuitable. The female *Ogyris* lays only after ants have touched her abdomen (Atsatt, 1981*b*). Other species like *Lycaena rubidus* drop their eggs on the ground where they are collected by ants (Funk, 1975).

Many myrmecophilous Lycaenids occur in association with Homoptera, and the ants tend both. Sometimes the caterpillars enter the shelters built by ants for Homoptera (Webster and Nielsen, 1984). They may feed on honeydew or prey on the Homoptera themselves. Other species like the Large Blue (*Maculinea arion*) are phytophagous for the first three larval instars. The fourth stage, however, stops feeding on *Thymus*, and drops to the ground. When an ant touches it the caterpillar adopts a curious humped posture and is carried into the nest by the ant. Inside the ant nest it feeds on ant brood. The ants appear to be neither hostile nor friendly to it in the nest, where it pupates and eventually emerges. For some reason only *Myrmica sabuleti* is a satisfactory host. Reduction in rabbit grazing in Southern England has produced a longer turf and lower soil temperatures. This has favoured *M. scabrinodis* over *M. sabuleti*, and led to the extinction of the Large Blue in this habitat (Thomas, 1984).

6.4 The cost–benefit balance in mutualism

In a genuine mutualism all parties must benefit from the association; it is clear from the earlier sections of this chapter that there are benefits to ants and to their various partners in the associations described. What is not so clear is the cost of the association to each participant. If the mutualism is advantageous to a partner, the benefits must be greater than this cost. While it has been argued that honeydew is a waste product of Homoptera and that its disposal by ants is a benefit not a cost, this may not be true if competition between honeydew producers leads to changes in the concentration or strength of honeydew. In any case Homoptera suffer costs, like predation by the ants that attend them, as well as changes in their population dynamics (e.g. reduced dispersal) which may themselves impose a reduction in fitness, that is, a cost. The cost situation is clearer with Lycaenid caterpillars, as they seem to have evolved a special gland whose secretion is more concentrated than the caterpillars' blood; this must entail a metabolic cost. Similar arguments can be applied to extrafloral and floral nectaries as well as other organs of myrmecophilous plants. These may have a cost of 10–20% of the plants' production. Protection of a caterpillar (or a plant) by ants must result in the production of more eggs (or seeds) than could have been produced if that much production had gone to further growth of the parent. It is difficult to measure the benefits where one or both partners are obligate mutualists, and it seems unlikely that, say a root-feeding aphid in a country where its winter host does not occur, has the option of withdrawing from the mutualism. It is still true, however, that if the benefits were in some way reduced the aphid would be under pressure to reduce its costs in order to retain a net gain from the association. The ants would also experience pressure to extract the highest benefit from the association, perhaps by eating aphids instead of honeydew, or to desert one mutualist for a more profitable one. Competition for mutualist partners should thus exist, and mutualisms can in no way be seen as cosy 'commitments' to assist a partner.

There is evidence that processes of this kind are at work in mutualisms. Addicott (1978*b*) found that the probability that a colony of *Aphis varians* would survive on a shoot of *Epilobium* was reduced if the shoot was close to bushes on which the same species of ant were tending the same or other aphids. He concluded that this was the result of competition for a limited amount of ant attendance. The limits could be set not by the number of ants but by their demand for the product, in this case sugars. Sudd and Sudd (1985) found that the response of *Formica lugubris* to sucrose solutions changed rapidly in the course of the season. In early spring workers in the nest had large reserves in the fat-body, but in this species these are rapidly used in the production of the sexual brood. Until midsummer the demand for sugar was high, and about 80% of ants tending aphids would accept 0.15 M sucrose. Ants at this time visited many species of aphids as well as the nectaries of bracken (*Pteridium aquilinum*). In late June the situation changed over a few days so that tending ants refused less-concentrated sucrose and only about 10% would accept 0.3 M sucrose. Bracken nectaries, and less ant-orientated aphids like *Cinara pinea*, *C. pilicornis* and *Thelaxes dryophilus*, were neglected or deserted, possibly because they could not compete with producers of better-quality honeydew. It looks as if there is more nectar and honeydew than the ants require, and the ants can therefore afford to neglect poorer quality. In this situation the cost of the mutualism to the aphis should increase: it will have to add substances to its honeydew which are of appreciable value to itself, or increase the quantity it produces or make it more easily available to ants, if it is to retain the ants' protection. It is now clear that the aphids ants tend are in any case not totally protected from predation by their guards, and that in summer they are a large part of the food fed to ant brood. Pontin (1978) enquires why aphids have not evolved some protection from this predation; many aphids have powerful chemical defences, but it is typical of obligate myrmecophiles that these are lost. In summer, it must be remembered, aphids reproduce exclusively by viviparous parthenogenesis: each vivipara is surrounded by a small group of daughters. In tended species where dispersal is reduced, these will form a clone of genetically identical individuals. If the price of ant attendance is the sacrifice of some of this clone to ant predation, there is nothing more

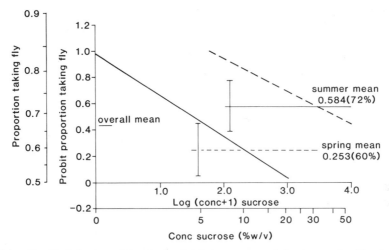

Figure 6.3 The behaviour of *Formica lugubris* foragers presented simultaneously with a dead fly and a drop of sucrose solution. Ordinate, the probit-transformed probability that the fly will be taken as prey; abcissa, sucrose concentration. In spring, predation is less likely, as the sucrose concentration is increased, and the results are well fitted by a straight line of negative slope. In summer, however, concentration has no effect, and the probability of capture can be treated as varying irregularly about a single mean value (after Sudd and Sudd, 1987).

extraordinary in this than the sacrifice of energy or protein which might have been used directly for aphid production (Sudd, 1987). It could be seen as a sort of social autotomy, analogous to the sacrifice of its tail by a lizard! Sudd and Sudd (1987) have found that in spring ants offered a drop of syrup and a dead fly simultaneously will accept the sugar if it is concentrated enough, and leave the fly. However, in summer they always take the fly, even if 2 M sucrose is offered as an alternative (Figure 6.5). Perhaps aphids have to pay in protein for protection in summer. The nitrogen metabolism of aphids is unusual in that they have no malpighian tubules and that many possess mycetomes, containing symbiotic micro-organisms, which may recycle excretory nitrogen. Possibly the cost of protein is not as high to them as it might seem.

A situation where some members of a clone 'accept' predation to increase the fitness of their identical sisters, and thus their own inclusive fitness, would be particularly open to exploitation by other insects. For instance, an unrelated insect, which lived among the mutualists but was more protected from predation, would enjoy the benefits of the mutualism, and since it would not be so heavily predated, it would increase its representation in the community. This would tend to destroy the mutualism, either by reducing tending or by increasing predation. It is worth considering the possibility that some lycaenids live in this way, however, since some caterpillars at least seem only to be tended by ants in the company of homopteran mutualists. At very least these must be a kind of Batesian mimic, sharing the cost of ant attendance with the Homoptera, and perhaps forming with them a larger honeydew source whose attendance is more worthwhile. If for any reason the caterpillars are relatively protected from predation, they would be obtaining protection under false pretences, rather as Müllerian mimics avoid attack without the cost of repellant secretions. Lastly, some seem to be aggressive mimics, obtaining protection from ants while preying on their Homoptera.

CHAPTER SEVEN

ANTS EXPLOITING ANTS

Perhaps the most sophisticated tools for dissecting the social structure of ants are those ants that exploit other ants. These exploiters range from ants that merely rob others to those that use common trail systems, while yet more devious forms share both the nests and the food supply of their hosts. Some species of ants are temporary parasites and produce normal workers which gradually take over the economy of established colonies of other species. There are also permanent parasites such as inquiline queens who behave as 'cuckoo ants', dispensing with a workforce altogether and using other colonies to raise their sexual offspring. Another form of permanent parasites are the slave-makers; these produce specialized fighting workers who kidnap the pupae of other species and use the adult workers that develop from this hostage brood as a slave workforce. All of these relationships involve opening up the communication systems of the exploited species. We will now explore how this is done.

7.1 Types of exploitation

7.1.1 *Mugger ants*

Some species of ants appear to behave as armed robbers, stopping foragers of other species and demanding with menaces that they hand over their prey items. Maschwitz and Mühlenberg (1973) observed up to three workers of the Sri Lankan *Camponotus rufoglaucus* lying in ambush at the nest entrance of a *C. sericeus* colony. When a homeward-bound forager approaches such an ambush site, the *C. rufoglauca* workers hurl themselves on their victim, who would typically not defend itself other than by rolling up into a ball. In so doing the mugged ant drops its prey item which is seized by the robbers and carried home. Oddly, the robber ants cannot distinguish between foragers with and without prey items, and attack them indiscriminately.

7.1.2 Claim-jumpers

Other species of ants are more subtle in the way they steal the food supply of the ants they exploit; they first exploit their highway. In Trinidad, workers of *Camponotus beebi* can be found using the chemical trails laid by the much more common *Azteca chartifex*. Workers of these two species rarely meet because the *Camponotus* workers take the day shift whilst the *Azteca* workers take the night shift (Wilson, 1965). Nevertheless, trail exploitation by *Camponotus* might be a form of parasitism if both species utilize the same food sources. In South American forests there is similar common trail use by *Crematogaster linata parabiotica* and *Monacis debilis* which live in close proximity to one another but do not share the same nest. Another *Crematogaster* species and *Camponotus fermoratus* may share a common trail without hostility; workers of the different species feed each other, and there is even mutual defence of their common nest (Wheeler, 1921; Weber, 1943). Could such association involve an interspecific division of labour? As yet no one knows: but this and similar systems are almost certain to reward further study. The remarkable point about this mutual trail-following symbiosis is that it commonly occurs between ants of different subfamilies, which have nevertheless apparently been able to read one another's trail pheromones and social signals.

7.1.3 Thief ants

While certain species simply mug foragers of other species for their prey items, others have become much more specialized thieves, entering foreign nests and stealing their brood. The European *Solenopsis* (*Diplorhoptrum*) *fugax* is such a specialist thief. In common with several other members of its subgenus it eats brood stolen from foreign ants' nests. *S. fugax* is a tiny ant, yet it preys on colonies with much larger workers such as *Formica fusca* and *Lasius flavus*. *S. fugax* workers use a distinct chemical strategy to break through the defences of their much larger foes. Scouts of *S. fugax* build complex systems of subterranean tunnels to lead them to the brood chambers of neighbouring prey colonies. When suitable brood is discovered the scouts use secretions from their Dufour's glands to lay recruitment trails back to their own nest. This summons large numbers of their nest mates who invade and raid the target nest (Hölldobler, 1968).

During the raid the *S. fugax* workers also deploy alkaloidal substances from their poison glands that repel the workers of the prey species. These substances are so repugnant and long-lasting that they effectively prevent the workers of the target nest defending their own larvae. The active repellent allomone is known to be highly effective against 18 different ant species (Hölldobler, 1973; Blum *et al.*, 1980). One particular thief ant has changed its habits and become a thief in human dwellings. The infamous Pharaoh's ant, *Monomorium pharaonis*, which is now a worldwide pest in hospitals and other public buildings, was originally a thief and scavenger ant in eastern India (Dumpert, 1981).

7.1.4 Guest ants

Some ants such as *Formicoxenus* have evolved into a much closer relationship with their host species than have *Diplorhoptrum* thief ants. *Formicoxenus nitidulus* is the only known guest ant species in Europe and spends its entire life in the large nests of wood ants. *Formicoxenus* is phylogenetically extremely close to *Leptothorax*, but this shiny guest ant has many unusual features. In addition to its burnished appearance, the female castes are poorly differentiated with many intermediate forms between workers and queens. However, the small colonies, which contain less than 100 individuals, have only one functional fertile female at any one time. The males of *Formicoxenus* are wingless, and resemble workers in external appearance. These odd features of both males and females of *Formicoxenus nitidulus* suggest that sexuals of this rather widely distributed and comparatively rare ant often mate with members of their own colony. Though *F. nitidulus* colonies maintain their own separate nest within the wood ants' nest they gain food from host workers either by actively stimulating regurgitation from the *F. rufa* workers, or by clinging on to the head of a worker as it feeds a nestmate and intercepting some of the food. The guest ants are rarely attacked by the hosts, and even then only rarely do these tiny shiny ants have to defend themselves with their powerful stings. In North America there are many other species of guest ant, particularly leptothoracines, who live with either *Formica* or *Myrmica* species. Guest ants exploit distantly related host species that are often much larger than themselves, and apparently do their hosts little harm. By contrast, all the remaining ant parasites considered in this chapter are closely related to their hosts and exploit them in a much more detrimental way, often curtailing the sexual production of their victims completely.

7.2 The temporary and permanent parasitic ants

The queens of temporary parasitic species avoid the hazards of independent colony foundation by entering another colony and diverting its economy to raise their own workers. Since the temporary parasite queen sooner or later kills her host queen(s) the number of host workers gradually falls because they are not being replaced. As this happens the workers of the parasite species take over the roles of the diminishing host worker population. Temporary parasite workers are similar in their morphology to free-living workers and must be able to accomplish all the foraging and household tasks because they will later raise the sexual offspring of their mother queen. The queen, however, is often specialized for infiltrating host colonies.

Slave-makers are similar to temporary parasites in many ways, indeed some species can be either slave-makers or temporary social parasites. The functional difference is that in slave-maker systems the host-worker population is continually supplemented as the slave-maker workers raid other host colonies to capture their brood. The new slave workers that emerge from this stolen brood work for their new colony mistresses, raising more slave-maker workers and sexuals. Slave-maker workers rarely need to perform household tasks and have therefore been largely free to evolve as specialist fighting machines for their role in slave-raids.

An inquiline queen, that is a queen of a parasite species that has dispensed with it own workers and uses others to raise its sexual offspring, may employ either of two rather different strategies towards the host queen. She may destroy the host queen in the colony she parasitizes, in which case the host workforce will steadily decline and must be used as quickly as possible to produce new parasite sexuals that can disperse to new colonies. Alternatively in some species the inquiline queen can live alongside the host queen, harnessing both her present and her future generations of host workers for the long-term production of parasitic sexuals.

Before we further examine temporary parasites, slave-makers and both types of inquiline, we consider their one common feature.

7.2.1 *Infiltration by parasitic queens*

With some notable exceptions (for example *Harpagoxenus* slave-makers: see below), parasitic queens all share one special feature; they are the only adult ants that can infiltrate established ant colonies of other species and be accepted by their workforces. Such queens must have either bypassed or cracked the colony-specific visa systems of their host colonies

Parasitic queens use a marvellous variety of techniques to do this. For example, the inquiline queen *Anergates atratulus* appears to play dead when met by a worker of the host species *Tetramorium caespitum*. She then clings with her jaws onto the worker's antenna and is dragged into the nest (Wheeler, 1910; Gösswald, 1954). *Epimyrma stumperi* either plays dead in a similar manner, or captures a *Leptothorax tuberum* host worker, rubs it with her specially modified forelegs, and then grooms herself, apparently transferring the host's colony odour to herself in this way (Kutter, 1969). Another parasite of *Tetramorium caespitum*, *Teleutomyrmex schneideri*, employs a third method of entry, and seems to appease the host workers with surface glandular secretions (Stumper, 1950; Gösswald, 1953). By contrast, *Leptothorax kutteri* employs a strategy of 'divide and rule'. She daubs the *L. acervorum* workers of the colony she is trying to infiltrate with a secretion from her huge Dufour's gland. This remarkable 'propaganda secretion' causes the *L. acervorum* workers to attack one another rather than the *L. kutteri* queen. The secretion can be passed from one host worker to another so that large numbers become contaminated until a substantial fraction of the worker population is engaged in deadly internecine battle (Figure 7.1).

Figure 7.1 The effect of the propaganda substances of the workerless parasite *Leptothorax kutteri*. The *Leptothorax acervorum* worker in the centre is being attacked by its own nestmates after being treated with the secretions of the Dufour's gland of *L. kutteri*. (Photograph by N. R. Franks.)

This may help to give the parasitic queen enough time to enter the colony and acquire its odour (Allies et al., 1986).

Queens of the slave-maker species *Harpagoxenus sublaevis*, like *L. kutteri* queens, have hypertrophied Dufour's glands that produce propaganda substances which also cause *L. acervorum* host workers to attack one another (Allies et al., 1986). However, in this case the *H. sublaevis* queen is not accepted by the workers of the host colony she enters. Instead, she deploys her propaganda substance not as a diversion but in an overt attack which enables her to demolish or drive off with her huge secateur-like mandibles all the adults of the colony she attacks. These slave-maker queens attack host colonies which have a large brood of worker pupae. The pupae quickly develop into adult *L. acervorum* workers who accept the slave-maker female as their queen. This first slave-workforce raises the first generation of slave-maker workers. Later these also use both their own propaganda substance and their fierce mandibles to raid *L. acervorum* colonies. There they steal worker pupae which eclose back in the slave-maker nest into yet more slaves.

Very little is known about how new queens in the formicine slave-makers, such as *Formica sanguinea* and *Polyergus*, enter host nests. However, what evidence there is suggests that these slave-maker queens are accepted by the existing workers and that unlike *Harpagoxenus* they do not have to drive away all the adults. It is possible that colony reproduction in *Polyergus* and *Formica sanguinea* is often associated with the temporary readoption of newly-mated queens into their maternal nest followed by the budding of such parental nests into daughter colonies (Marlin, 1968). When newly mated queens of the American amazon ant *Polyergus lucidus* were presented, in the laboratory, with small artificial colonies of workers of either one or other of their two host species, they were more succesful at being adopted into the colony with the same species of slave as their maternal nest (Goodloe and Sanwald, 1985). This suggests that the odours of the slaves may also affect the slave-makers that they raise. This may lead to host choices being reinforced from one generation to the next.

Having emphasized the importance of non-independent colony founding in all these social parasites, we now consider other aspects of their behaviour and natural history.

7.2.2 Temporary parasites

A surprising number of species of ant colony begin life as temporary parasites: these include many wood ants of the *Formica rufa* group (Creighton, 1955) and certain *Lasius* species.

The majority, if not all, of new nests formed by colonies of species belonging to the *F. rufa* group of wood ants, such as *F. lugubris* and *F. polyctena*, are created through the readoption of newly-mated queens by their maternal or possibly another conspecific nest and the subsequent budding of the established nest. However, some newly-mated queens of *F. rufa* group do not return to their parental nest but enter nests of species in the *F. fusca* group such as *F. fusca* and *F. lemani*. If the *F. rufa* queen is successful in entering a *F. fusca* nest the host queen or queens are eliminated, in some as yet unknown way, and the *F. rufa* group workers are raised to form a 'mixed nest'. Eventually as the *F. fusca* workers die the colony becomes purely a wood-ant nest.

F. rufa queens seem to have no special chemical, morphological or behavioural adaptations for the dangerous task of entering alien *F. fusca* colonies. Perhaps the reason *F. rufa* group queens have adopted this apparently supremely hazardous way to start a colony is that the alternative of claustral colony foundation would be even less likely to succeed. A tiny wood-ant nest founded independently of either its parent colony or a interspecific host would soon be overrun by larger alien conspecific nests. Indeed, such large nests sometimes stage wars against one another (Mabelis, 1979). So by hiding herself and her initially small worker population in a *F. fusca* colony, a *F. rufa* queen may have at least some chance of survival. Thus intense intraspecific competition may have been one of the major selection pressures that promoted the evolution of temporary parasitism in wood ants. As we will discuss later, intraspecific competition may also have played a key role in the evolution of the many different forms of permanent parasitism in ants.

In the formicine genus *Lasius*, parasite queens are much smaller than congeneric queens that found their nests independently. *L. fuliginosus*, which occurs in southern England, is believed to be an obligate parasite of *L. umbratus* nest which oddly enough are also temporary parasites—in this case of *Lasius niger*. It is believed that all temporary parasites in *Lasius* destroy the host queen, but the methods they use are largely unknown. However, *L. reginae* queens are known to murder their much larger *L. alienus* host queen by first overturning her and then taking a stranglehold on her neck until they have choked the life out of her (Faber, 1967).

7.3 The evolution of inquilines

We now consider the evolution of workerless parasites. Queens of this type of permanent parasite utilize the established economy of host colonies to raise their purely sexual offspring.

One of the most remarkable things about such inquilines, and indeed the other type of permanent parasite, slave-makers, is that in almost every case the host species is the closest living relative of the parasite. The parasites and their hosts are like sibling species sharing the same recent ancestor. Such is the universality of the close phylogenetic pairing of exploiter and exploited that this relationship has been called Emery's rule, after the great ant taxonomist Carlo Emery, who first drew attention to it. All permanent parasites must rely purely on the host workers to raise their sexual offspring and to realize their fitness. For this reason their communication with and control over the host colony must be very precise. For example, when a permanent parasite dispenses with the host queen, the host workers, freed from the so-called queen-control emanating from their own mother, might be expected to switch their colony's economy entirely to the production of their own sons. Clearly the parasite must control the life of her host colony with pheromones similar to those produced by the normal queen. This would explain why parasites have to be so closely related to their hosts.

The close phylogenetic relationship of host and parasite suggests that host species may have given rise to their own parasites. Though apparently far-fetched, this idea is not too hard to envisage given the conflicts that occur within social insect colonies. In Chapter 1 it was suggested that these conflicts are so strong that the secondarily accepted queens in a polygynous nest of *Myrmica rubra* could be considered as intraspecific parasites of the queens that started the colony (Elmes, 1973). If such invasive queens became reproductively isolated from their hosts, as might occur, for example, if they had their nuptial flights at a different time of year, then a new parasite species could evolve. This scheme must be accepted with caution, however, because it invokes sympatric speciation which is believed to be much rarer than allopatric speciation (Wilson, 1971).

Once the parasites are reproductively isolated from their hosts they appear to become progressively miniaturized through evolutionary time. The fitness of parasite queens will depend largely on their production of a very large number of sexual offspring. If they used the host workers to raise smaller rather than larger daughter queens more could be produced, and small queens would not be penalized by the lack of body resources as they would not have to found new colonies independently. The next step would be for the parasites to dispense with their own worker production using all the resources of the host workers to raise their sexual offspring.

Many elements of this scheme seem to occur in three species of *Myrmica*: *M. rubra*, *M. ruginodis* and *M. sabuleti*, all of which have colonies in which large and small queens can be found. In fact there has been a long-running debate about whether these microgyne and macrogyne forms are separate species of parasite and host respectively. Brian and Brian (1955) suggested that this was the case in *M. ruginodis* (originally referred to as *M. rubra* by these authors). Recently Elmes (1978) has shown that the small queen form in *M. sabuleti* is a separate species which is now called *M. hirsuta*. Using electrophoretic techniques, Pearson and Child (1980) have presented evidence suggesting that the macrogyne and microgyne forms in *M. rubra* are reproductively isolated. The electrophoretic technique when applied to the worker populations in nests with the two types of queen suggests that the microgyne forms usually produce no workers (Pearson, 1981). In rearing experiments however, Elmes (1976) showed that the microgyne form was at least still capable of producing worker forms. Thus the microgyne forms seem not to have crossed the threshold of becoming obligate workerless parasites. The microgyne form in this *M. rubra* complex seems to specialize in massive production of sexual offspring. The number of sexual daughters produced per queen is 37 times higher for the microgyne form than the macrogyne form. In addition the microgynes seem to produce their own sons whereas in the macrogyne form it is the workers who produce most, if not all, of the males (Pearson, 1981; Smeeton, 1981).

One further advantage of miniaturization for the microgyne forms could be that they masquerade as workers rather than sexuals during the larval phase. In *Myrmica rubra* workers seem to control which larva develop into queens or workers and they prevent too many sexual females being produced by attacking them so that they metamorphose prematurely into workers (see Chapter 5). By producing small female larvae the microgyne form is possibly avoiding this sterilization procedure.

Distinct large and small queens also occur in the North American *Leptothorax longispinosus*. The tribe Leptothoracini probably has a greater ratio of parasitic to free-living species than any other group of ants. Though the species status of the macrogyne and microgyne forms of *L. longispinosus* is unknown, their presence in one of the many *Leptothorax* species may point to a similar evolutionary pathway to inquilinism as in *Myrmica*.

Three species of workerless inquiline parasitize *L. acervorum* colonies in Europe. In order of increasing parasitic specialization these are *L. gosswaldi*, *L. kutteri* and *Doronomyrmex pacis*. Like most other inquilines these species are extremely rare. Queens of *L. gösswaldi* can only be distinguished from *L. acervorum* queens by the smallest differences in morphology, and are probably exceptionally closely related to their hosts. *L. gosswaldi* queens may only be able to enter queenless host colonies, which they will therefore only be able to exploit for a short time to raise their sexual offspring before the host workers die off. *L. kutteri* is a more specialized social parasite than *L. gosswaldi*. *L. kutteri* females, though again very similar to their hosts, are smaller than *L. acervorum* queens, being about the same size as *L. acervorum* workers. These parasitic queens are more common in polygynous than monogynous host nests where they live peacefully alongside the host queens which continue to lay eggs. Nevertheless, *L. kutteri* queens appear to have to fight their way into new host colonies using the special propaganda substances mentioned in Section 7.2.1. Once established, *L. kutteri* queens groom host queens at an unusually high rate, possibly to gain the colony odour of their host nest and thereby remain undetected. Observations show that *L. kutteri* queens suffer no aggression from any member of the host colony once they are established. This is remarkable because they even eat some of the eggs of the host queens.

Doronomyrmex pacis females also live alongside their host queens. Indeed, *L. acervorum* colonies parasitized by either *L. kutteri* or *D. pacis* continue to produce their own sexual offspring and workers in addition to the males and females of the parasite species. *D. pacis* shows an even greater tendency to parasitic specialization than *L. kutteri*; it is even smaller and has the typical hairless and shiny appearance of the most advanced inquilines. The close phylogenetic affinity of these three parasites and their host has been beautifully demonstrated by similarities in their pheromones (Buschinger, 1972, 1975). Virgin queens of all three parasites call males by releasing from the tip of their sting sexually attractive pheromones secreted in the poison glands (Chapter 5). The sexual pheromones of the three parasitic species are attractive not only to their conspecific males but also to males of some of the other species. This is exceedingly odd because such substances are usually species-specific, and often act as isolating mechanisms. Even more surprising, in certain cases, hybrid progeny resulted from these interspecific matings. This suggests that these three parasite species are exceptionally closely related.

7.3.1 The ultimate cuckoo ants

Even more specialized inquilines occur in other groups of ants. We will conclude this consideration of the evolution of these forms by looking at two parasites of *Tetramorium* pavement ants.

The first of these is the parasite *Anergates atratulus*. Virgin females of this species are tiny compared with the size of the *Tetramorium caespitum* host queen, indeed they are about the same size as a *Tetramorium* worker.

However, on entering a host colony an *Anergates* female rapidly becomes much bigger, and, as it devotes all its energy to egg production, its gaster swells to many times its original size. *Anergates* males are even more specialized than the females; they too are tiny but are wingless, and have a barely sclerotized cuticle so that they look like walking pupae. The males inseminate their virgin sisters before most of the latter fly away to try to enter other *Tetramorium* colonies (Figure 7.2). Since *Anergates* females are never found alongside host queens, but probably could not kill the much larger host queens, it is believed that they can only enter queenless *Tetramorium* colonies. This might account for the great rarity of this species compared to the abundance of its host. Nevertheless, *Anergates* has now become established in North America after the accidental introduction of its host from Europe. The tiny size of *Anergates* females means that they can be produced in huge numbers by the host colony: male production is probably reduced so that an established *Anergates* female probably produces only sufficient males to inseminate all her daughters. In this way, the greatest number of new *Anergates* females will be produced before they embark on the extremely hazardous task of finding a new queenless host nest.

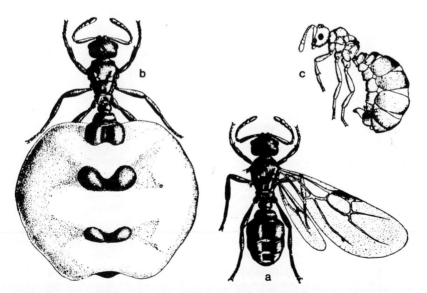

Figure 7.2 The parasite *Anergates atratulus*: the workerless females live in colonies of *Tetramorium caespitum*. (*a*) Virgin queen; (*b*) old queen with physogastric abdomen; (*c*) wingless pupa-like male. (Redrawn from Wheeler, 1910.)

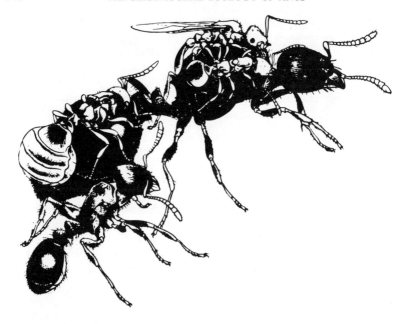

Figure 7.3 Females of the ectoparasite *Teleutomyrmex schneideri* riding on their *Tetramorium caespitum* host queen. A *Tetramorium* worker stands alongside its queen. One of the parasitic females still has wings and is probably a virgin. The older mated parasite on the host's abdomen is physogastric with eggs. (Redrawn from Wilson, 1971, after a painting by W. Linsenmaier.)

The oddly-named *Teleutomyrmex schneideri* (the name means 'final' or 'ultimate' ant) is also a parasite of *Tetramorium caespitum*. The only ultimate thing about *Teleutomyrmex* is its ability to get other ants to do all the work. *Teleutomyrmex* females, like *Anergates*, have completely lost their own workers and they are even smaller than *Anergates*. *Teleutomyrmex* females are actually ectoparasites who ride on the host queen (Figure 7.3). Older *Teleutomyrmex* females ride on the gaster of the host queen so that their physogastric abdomen lies above that of their host. In this way the parasite inserts its own eggs into the brood production line of the host colony. In common with many other workerless parasites, *T. schneideri* has an enigmatic distribution. It was first discovered by Heinrich Kutter in the high Alps near Zermatt and later rediscovered at Briançon by Cedric Collingwood also at a height of 2000 m (see also Buschinger, 1985).

7.4 Slavery

We have left the discussion of slavery in ants to the end of this chapter simply because its evolutionary origin is so delightfully controversial. Myrmecologists, and indeed Darwin himself, have argued about the evolution of the slave-making habit ever since the behaviour was first discovered by Pierre Huber in 1804. We now know that the habit of slave-making in ants has evolved independently at least four times, twice in the Formicinae (in the *Formica, Polyergus, Rossomyrmex* group and in *Myrmecocystus*) and twice in the Myrmicinae (in the Leptothoracini— *Harpagoxenus, Leptothorax duloticus, Epimyrma* assemblage and in *Strongylognathus* which exploit *Tetramorium* species) (Alloway, 1980).

To provide some further background on the extraordinary habits of slave-making ants we briefly review the natural history first of formicine and second of myrmicine slave-makers. Examples are drawn mostly from the two best-known genera: *Polyergus*, which raids *Formica* nests, and *Harpagoxenus*, which enslaves various *Leptothorax* species.

7.4.1 *Amazon ants*

There are three palaearctic and several nearctic species of Polyergus. The best known are the European *P. arufescens* (Figure 7.4), as first described by Huber (1810), and the North American *P. lucidus* and *P. breviceps*. The scale of raiding by all three of these species is otherwise seen only in army ants.

P. breviceps can recruit 2000 ants into a single raid against a host colony which may be 75 m from the slave-makers' nest (Topoff *et al.*, 1984). Forel (1874) intensively studied a single *P. rufescens* colony in Switzerland. During 33 days he saw this colony make 44 expeditions and estimated that the ants retrieved a total of 40 000 larvae and pupae from the various *F. fusca* and *F. rufibarbis* nests they plundered. Recently, Topoff *et al.*, (1984) have made a study of the behaviour of *P. breviceps* in the high desert of Arizona. A raid typically begins when a scouting *Polyergus* worker finds a host *Formica* colony at some distance from its own nest. It then returns home, orientating purely by optical cues and without laying a trail. Once home the scout excites its *Polyergus* nestmates and leads the raid back to the target colony. On the outward raid the *Polyergus* scout orientates again by the position of the sun and polarized light patterns in the sky. However, as it proceeds the scout lays a recruitment trail which its nestmates follow: they also lay a trail so that the route to the *Formica* colony is well marked. If the scout hesitates, the ants in the *Polyergus* swarm front

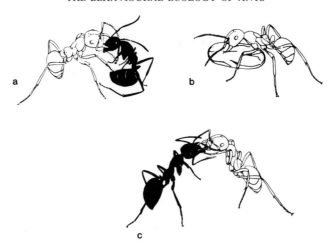

Figure 7.4 A cartoon of some of the major activities of *Polyergus rufescens*, slave-maker workers. (*a*) The stereotyped attack of a slave-maker on a *Formica* worker during a slave-raid. The slave-maker uses its sickle-shaped mandibles to pierce the brain of the *Formica* worker. (*b*) The capture and transport of a Formica worker pupa. (*c*) *Polyergus* being fed by a *Formica* slave. (Redrawn from Wilson, 1971, after a drawing by Turid Hölldobler.)

begin moving in expanding circles until the scout again takes the lead or one of them finds the target nest. In this way, if the scout becomes lost the raid still has a good chance of finding its target.

When they hit the target colony the real fight begins: the defending workers try to wall up the entrances of their nest, whilst the amazon ants try to break through. As the latter do so they engage the defenders one-to-one. The *Polyergus* workers are more strongly built than the *Formica* defenders and have sabre-shaped, toothless and very sharp mandibles that they use to pierce the brains of defending workers (Figure 7.4). After the raid *P. breviceps* workers use both optical and chemical cues to return to their own nest, with the brood they have kidnapped.

Two other North American formicine slave-makers, *F. pergandei* and *F. subintegra* (Figure 7.5) produce large quantities of acetates from their hugely hypertrophied Dufour's glands. These acetates mimic the compound undecane that their host workers would normally use as an alarm pheromone. However, such is the strength and volume of the slave-makers' secretions that they cause host workers to fly into a complete panic. Not only do these propaganda substances make the target colony much easier to raid, they also serve as an additional recruitment signal for the slave-

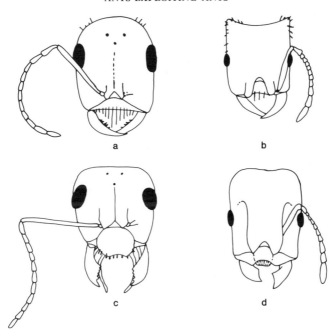

Figure 7.5 A rogues' gallery of slave-making ants, showing the partially convergent evolution of the fighting adaptations in four species: (a) *Polyergus rufescens*, (b) *Strongylognathus testaceus*, (c) *Formica sanguinae*, (d) *Harpagoxenus sublaevis*. *Polyergus* and *Strongylognathus* have piercing mandibles. *Harpagoxenus* has secateur-like mandibles to snip off the limbs of its opponents. *Formica sanguinae*, by contrast, is a facultative slave-maker and has unmodified mandibles. (Redrawn from Kutter, 1969.)

makers (Regnier and Wilson, 1971). Similar substances probably also occur in *Polyergus* and may account for Huber's (1804) observations of the reluctance of the *Formica* workers to return to their own nest following a raid.

7.4.2 The tiny slave-makers

In contrast to the massive scale of operations in *Polyergus*, slave-makers such as *Harpagoxenus americanus* live out their entire lives on a stage not much bigger than a large table-top. In broad-leaved woodland across most of the eastern United States and southern Canada, *H. americanus* can be found enslaving *Leptothorax curvispinosus*, *L. longispinosus* and *L. ambiguus*. Both slave-maker and host species commonly nest in hollow acorns, with more than 200 ants in the same nut. These tiny ants quickly take to living in the laboratory where their behaviour and slave-raids can be observed in detail.

In *H. americanus* individual scouts find host colonies near their own nest and return home only when they have ascertained the exact whereabouts of the entrance to the target colony. Scouting in *H. americanus* is remarkably similar to that in *Polyergus*, but in miniature. When the *H. americanus* scout does return home she runs around within the nest in an extremely agitated state, apparently using first mechanical and then chemical signals to whip up enthusiasm for the raid among her nestmates. When she leaves the colony she goes in a straight line to the target nest, dragging her protruded sting on the substrate to lay a recruitment trail which may lead both slave-makers and slaves from her nest to the target. In this way a tight single-file formation is led into battle.

The *H. americanus* workers run round the defending *Leptothorax* workers, nipping them with their secateur-like mandibles, and can quickly snip off the legs and antennae of the defenders. *H. americanus* may also use propaganda substances, as there appears to be widespread panic in the host colonies they attack (Wesson, 1939; Alloway, 1979). Once they have fully taken possession of the target nest, both the slave-makers and their slaves begin to transfer the brood home, concentrating on the worker pupae and large larvae.

The European *H. sublaevis*, which occurs in Germany and the Alps and throughout much of Scandinavia, shares many similarities with *H. americanus*. This is particularly intriguing as these two slave-makers enslave different phyologenetic groups of leptothoracines and are believed to have undergone an independent and convergent evolution (Buschinger, 1981).

H. sublaevis lives most commonly in twigs on the floor of dry pine woods, where it enslaves *Leptothorax acervorum*, *L. muscorum* and *L. gredleri*. *H. sublaevis* uses rather different raiding tactics from *H. americanus*. Individual scouts find potential target colonies and return home just as in *H. americanus*, but the scout individually leads one slave-maker worker at a time into battle by tandem-running (see Chapter 5). This is a much more primitive form of recruitment than the trail-laying used by *H. americanus*, and results in a much slower build-up of attackers at the target nest. The Canadian *H. canadensis* is believed to be quite closely related to *H. sublaevis* but has an even more primitive form of recruitment. At the beginning of a slave-raid *H. canadensis* workers also employ tandem recruitment, but will do so immediately after finding alien host workers and without first discovering the nest entrance of a target host colony (Stuart and Alloway, 1982). The various *Harpagoxenus* species also use somewhat different fighting techniques. Often *H. sublaevis* workers allow defending *Leptothorax* workers to take hold of each of their legs and antennae. The *H. sublaevis* rarely come to harm when they are pinned down in this way, and effectively they can take up to eight defenders out of the battle, whilst other slave-makers attack the host nest and capture its brood. The different, and arguably more primitive, recruitment and fighting techniques employed by *H. sublaevis*, in comparison with *H. americanus*, are probably one of the factors that have shaped the different slave-maker/slave ratios in these two species. In *H. sublaevis* there are often twice as many slave-makers per slave as in *H. americanus* colonies.

In addition it has recently been shown that *H. sublaevis* workers, like their colony-founding queens, smear special chemicals from their massively hypertrophied Dufour's glands onto defending host workers. These chemicals, like those used by *L. kutteri* inquiline queens when they also try to infiltrate *L. acervorum* colonies (see Section 7.1), have the remarkable effect of causing the defending workers to attack one another. The occurrence of disruptive propaganda substances, in both *Harpagoxenus* and *Formica subintergra* and the specialization of the mandibles for fighting in *Harpagoxenus* and *Polyergus*, are a remarkable illustration of convergent evolution in these completely separate lineages of slave-making ants.

There are other slave-makers that utilize various leptothoracine species, including the North American *Leptothorax duloticus* and the European *Epimyrma*. *L. duloticus* is a particularly interesting species because it can use all three of the same slave species that *H. americanus* exploits. Scouting and recruitment are based on much the same techniques in *L. duloticus* and *H. americanus*, but the former has a very different fighting technique. *L. duloticus* has unspecialized mandibles much the same as its host species, and its slave-raiding weapon is its sting. *L. duloticus* has a very large poison gland relative to its body size, and fights defending host workers by stinging them in the mouth. The poison it injects is extremely potent and causes the death of the defender in a matter of seconds. Certain *Epimyrma* species utilize a similar stinging technique when they attack their *Leptothorax* (*Temnothorax*) host colonies (Buschinger *et al.*, 1980).

7.4.3 Imprinting and slave-making

In the enigmatic world of ants it should perhaps come as no surprise that the slave-making habit is only possible through the highly tolerant behaviour of certain individuals. First, the slave-makers and their slaves must tolerate the new brood they capture and allow it to develop and eclose rather than eating it. Second, the eclosing slave workers must tolerate, recognize, accept and work for their captors' colony as opposed to any other. This last point is particularly intriguing as the slave-maker colony may continue to live next door to the colony from which it obtained its slaves, yet such slaves do not return home except in species where they help their captors in further raids against their own parental nest!

Generally ants seem to learn to recognize the kind of brood that was present in the nest they developed in, and will tolerate and serve this brood in preference even to that of their own species (see Chapter 5). Indeed, slaves must recognize and rear brood, not only of the slave-makers but of potential new slaves, which may have been captured from many foreign nests of their own and sometimes also of two or three other species. Larvae, pupae and callows have less distinct colony-specific visas than any other individuals in ant nests (see Chapter 5) and may be transferred between nests more easily than any other forms. Young, newly eclosed callow workers also may learn to recognize the colony-specific odour of their nest mates. After a learning period which occurs in the first two weeks of adult life, workers of certain species will thereafter treat all nestmates as friends and all others (including full sisters reared elsewhere) as foes. Such critical-period learning, analogous to imprinting in ducks, has been well documented in such formicine ants as *F. sanguinea* and *F. cunicularia*, one of its slaves (Le Moli and Mori, 1985). Clearly the slave-makers must imprint on their slaves (and vice versa) for them to have a good working relationship. This might explain why *Polyergus* queens seem to choose new host colonies of the same species as the slaves who reared them (see 7.5).

7.4.4 *The evolution of slave-making*

A number of theories have been put forward to explain the evolution of slave-making in ants. Darwin (1859) suggested that slavery arose from one type of colony preying on another to capture and eat its brood. If a lot of brood was captured and some eclosed as adult workers before they could be eaten, such workers might become incorporated in the social structure of their captors' colony. In this way a behaviour pattern that first involved predation could become adapted for slave-raiding. Wilson (1971) put forward a modification of this theory suggesting that the ancestor of a slave-maker might not have been a specialist predator of other ants but might have been a highly territorial species. Established colonies of many types are known to overrun smaller ones of the same or closely related species that infringe on their territory and compete for the same resources (Chapter 8). If a larger colony in the process of destroying a smaller one captured some of its brood, again initially for food, then a scheme like Darwin's might allow the further evolution of slavery from this form of territoriality. The advantage of Wilson's hypothesis is that it is better able to explain why slave-makers are almost always very closely related to their host species. Virulent territoriality is more likely to occur between closely related species who would compete for similar resources, whereas predaceous ants normally raid distantly related species.

There is considerable circumstantial support for the Wilson–Darwin scheme for the origin of slave-making. Support for the territoriality theory of slave-making comes from observations in the laboratory that large colonies of *Leptothorax curvispinosus*, a common North American slave of *L. duloticus* and *Harpagoxenus americanus*, will overrun neighbouring smaller colonies of the same species and later make 'slaves' of some of the captured brood (Wilson, 1975a). In extraordinarily large field collections of *L. longispinosus* and *L. ambiguus* colonies that also share the same slave-makers as *L. curvispinosus*, Alloway (1980) found a small number of mixed-species nests, which would be difficult to explain unless colonies of these normally free-living species occasionally raid one another and make slaves. Of course, colonies which were intraspecific slave-makers would go unrecognized as such in field collections because the slave-makers and slaves would be indistinguishable. The conjecture that territoriality is associated with slave-raiding is supported by intraspecific and interspecific 'slave-raiding' experiments with these species in the laboratory. Territorial raiding in *L. curvispinosus* has many features in common with the raiding behaviour of *L. duloticus* (Wilson, 1975a) in terms of the behaviour of the workers and their recruitment and fighting. In addition, *L. duloticus* is known to eat some of the smaller brood items it captures on a slave-raid (Alloway, 1979). In formicine slave-makers such as *Polyergus rufescens, P. lucidus* and *Formica sanguinea* the slave-makers are known to devour some of the brood they capture on slave-raids (Dobrzanska, 1978; Kwait and Topoff, 1983).

The fact that *H. canadensis* scouts may recruit to the territory of the target nest and not directly to its entrance also supports the hypothesis that slave-raiding evolved in some way from territorial conflicts (Stuart and Alloway, 1983; Stuart, 1984).

In addition, in the North American honey-pot ant, *Myrmecocystus mimicus*, territorial disputes have been observed in the field to turn into slave-raids when one colony greatly outnumbers the other (Hölldobler,

1976*a*) (see also Chapter 8). Similar intraspecific territoriality may also occur in the European *Tetramorium caespitum* (Alloway, 1980), which is also a host to slave-makers, and might, according to Emery's rule, be expected to show the precursors of their behaviour patterns.

Thus in present-day host species and in slave-makers there are numerous examples of behaviour patterns that might be expected to occur if the Wilson–Darwin hypothesis for the origin of slave-making had actually occurred. There is, however, one entire aspect of the behaviour of most slave-making species that the Wilson–Darwin theory does little to explain. The vast majority, if not all, of known slave-making species found their colonies non-independently. That is, slave-maker queens never start their colonies alone, but first infiltrate a colony of the host species before producing any offspring of their own. In this aspect of their lives, slave-maker queens are remarkably similar to the temporary parasites, inquilines and cuckoo ants that we have discussed earlier in this chapter. Indeed, this similarity between slave-makers and other social parasites is emphasized by the fact that slave-makers most often occur in the same groups of ants that also have the other forms of social parasite (Wilson, 1971).

Wheeler (1910) recognized the phylogenetic associations between the various forms of social parasite and suggested that some slave-makers might have evolved from temporary parasites. When the host queen is killed by a temporary parasite queen the host workers gradually die out and are replaced by the temporary parasite workers. However, if the temporary parasite workers raided other free-living host colonies, for either predatory or territorial reasons, then the temporary parasite might evolve along the lines originally proposed by Darwin (1859) and become a permanent social parasite combining the two crucial features of slave-making species—non-independent colony foundation and slave-raiding. *Formica sanguinea* is a prime candidate for such an origin of the slave-making habit as in certain parts of its range (e.g. England) it is almost always purely a temporary parasite whilst in other habitats (e.g. Switzerland) it is a fully-fledged slave-maker (see Darwin, 1859). The temporary parasite route to slave-making may therefore have been particularly important in *Formica*.

One possible reason for a correlation between non-independent colony foundation (and therefore temporary parasitism) and virulent territoriality (and therefore incipient slave-raiding), is that both are most likely to occur in habitats and in species in which competition is particularly intense. In such situations there would be strong selection both for non-independent colony foundation, in which new queens hide their growing colonies within established ones, and for territorial destruction of neighbours, in which large colonies wipe out smaller ones.

The temporary parasite route for the origin of slave-making is unlikely to have applied to all slave-making ants. For one thing, temporary parasites are unknown in leptothoracines. However, in the *Leptothorax* species that are hosts to slave-makers, and therefore according to Emery's rule believed to be close to the ancestors of their own slave-makers, polygyny and polydomy are unusually common. The existence of multiple queens and multiple nests in these host species has recently stimulated theories (Buschinger, 1970; Alloway, 1980) for the origin of slavery in the leptothoracines. According to Buschinger (1970), in polygynous and polydomous species queens might arise that are effectively specialized as parasites of the efforts of the other queens in the colony (see for comparison the natural history of microgyne *Myrmica* queens: Section 7.3). Such parasite queens might then produce fewer of their own workers. This could be a particularly good strategy in polydomous colonies because brood is transferred from nest to nest within the colony. The few workers produced by the incipient parasitic queen might then be able to bias brood transport between the subnests of the colony so that their own queen was served by more than her share of the workers produced by the other queens. The parasite queen could thus produce a very high proportion of all the sexuals raised in the polydomous and polygynous colony. This scheme has the advantage of explaining the close phylogenetic affinity between slave-makers and hosts, their non-independent colony foundation and the similarity between slave-raiding behaviour and normal brood transport behaviour in the host species. However, like certain scenarios for the evolution of inquilines (see Section 7.3) this scheme implies sympatric speciation: i.e. the incipient parasite would have to become reproductively isolated from its host whilst sharing the same habitat. Another possible objection is that not all modern host species, even in the leptothoracines, are both polygynous and polydomous. Alloway (1980) has incorporated the key elements of Buschinger's hypothesis, non-independent colony foundation associated with polygyny, into a theory largely based on territoriality. Suppose that two closely related and typically polygynous species were allopatrically isolated and one showed intraspecific territorial raiding and incipient slave-making, and that the species ranges of the two species came to overlap due, for example, to some climatic change. Then one species might be preadapted to both slave-making habits; non-independent colony foundation by virtue of its own polygyny and that of its perspective host, and slave-raiding stemming from its intraspecific territorial raiding.

7.4.5 Are slave-makers degenerate?

There has been almost as much debate about the direction of evolutionary transitions between the various forms of social parasitic ants as about the initial origin of these exploitative symbionts. As we have seen, Wheeler (1910) suggested that formicine slave-makers may have evolved from temporary parasites. Since modern slave-makers show both slave-raiding and non-independent colony foundation it is intriguing to consider which of these habits came first. Did certain slave-makers evolve from inquilines without slave-raiding workers, or have workerless inquilines ever evolved from slave-makers.

In certain parasitic genera, for example *Strongylognathus* and *Epimyrma*, there are apparently some species that are true slave-makers and others that rarely ever slave-raid. It is to these genera that we should look for possible transitions between the various forms of parasitism. *Strongylognathus testaceus* is supposed to coexist with its host *Tetramorium caespitum* without making slave-raids, whilst other *Strongylognathus* species have been observed to conduct extensive raids (Kutter, 1969).

The best-studied genus in which different and possibly transitional forms of parasitism occur is *Epimyrma* (Buschinger and Winter, 1983). Colonies of *E. ravouxi*, for example, have on average 25 slave-maker workers who actively participate in slave-raids to maintain a slave-population of about 140 *Leptothorax unifasciatus* workers. By contrast, in *E. kraussei* the colonies which have any *Epimyrma* workers at all contain on average only 4, and more than 40% of all parasite colonies have no parasitic workers at all. *E. kraussei* colonies have on average about 30 *Leptothorax* (*Temnothorax*) *recedens* host workers. *E. kraussei* colonies with parasite workers were capable of successful slave raids in the laboratory. Thus it is likely that *E. kraussei* is only a facultative slave-raider and rarely raids.

Epimyrma queens, unlike *Harpagoxenus* queens, can enter established host nests and come to be accepted by the already adult host workers. *Epimyrma* queens kill only the host queen but *Harpagoxenus* females must drive off or kill all the adults in the colonies they invade. Perhaps because they can enter and take over established colonies, the highest numbers of host species workers in nests of both of these *Epimyrma* species are close to the maximum number of workers in free-living colonies of the respective host species. This last point is probably most important of all in a comparison of the social systems and behaviour of *E. ravouxi* and *E. kraussei*. In both cases it is the host workers that are responsible for realizing the parasitic queens' fitness by raising her sexual offspring.

Parasitic workers, who rarely tend brood or bring food into the colony, are purely a means to the end of maintaining the host worker population. If the host workers they bring in are redundant, or if they cost more resources than they can replace, then the *Epimyrma* queen should dispense with producing any parasitic workers at all.

Thus the behaviour of an *E. kraussei* queen in essentially that of an *r* strategist (see Section 1.4.1). Such queens can apparently begin life with a full complement of host workers in their nest. In this case the queen's optimum life-history strategy seems to be to produce only sexual offspring. This means that the host workers will die out over a period of 2–3 years (Buschinger and Winter, 1983) but in this short time the queen may have been able to produce more sexual offspring than if she had 'wasted' some resources producing slave-raiding workers. Perhaps *E. kraussei* queens only produce workers if they have invaded very small host colonies and perhaps only then is this species a facultative slave-maker. By contrast, *E. ravouxi* is a *K* strategist producing many slave-making workers who maintain a host workforce for 10 years (Buschinger and Winter, 1983) so that sexuals can be produced over an extended period. Queens of certain other *Epimyrma* species seem to be obligate *r* strategists producing no workers of their own at all.

Species such as *E. kraussei* are often referred to as degenerate slave-makers (see for example Buschinger and Winter, 1983): however, this term can often be misleading as the various strategies used by such parasites are probably simply those that have the best chance of maximizing the queen's fitness, whether they involve producing large numbers of slave-maker workers, facultative slave-raiding or dispensing with worker production completely. For this reason it is difficult to determine first, which type of parasite, if any, is closer to the ancestral form or forms, and second, what transitions, if any, have occurred between the various parasitic forms.

The different strategies adopted by *Epimyrma* queens are only possible because they can infiltrate established host colonies and come to be accepted by the existing adult workers. The parasitic queens of *Harpagoxenus* and *Leptothorax duloticus* must first drive off all the adults in the host nest they invade. Only newly-eclosed workers will accept these slave-makers as nestmates. *H. americanus* colonies produce fewer slave-maker workers than *H. sublaevis*, and for this reason the American species has been thought of as being particularly degenerate (see Creighton, 1950). This view is untenable in evolutionary terms: simply, the more efficient each slave-maker worker the fewer should be produced by their colony. Recall that a *Harpagoxenus* queen during colony foundation can execute the equivalent of a slave-raid all by herself. *H. americanus* workers are almost the same size as queens and are almost as efficient at solo raiding.

Slave-maker workers should simply be regarded as a specialist caste in the economy of their colony. Just as with other specialist castes in monospecific colonies, as their efficiency increases then their relative numbers are reduced (see Chapter 4 on the economics of the division of labour in ants). Efficiency is also generally correlated with specialization and a reduction in the number of tasks performed. Thus 'ergonomic principles' that explain the role of specialist castes in free-living colonies also help to explain the organization of slave-maker nests. Highly specialized slave-maker workers occur in relatively low numbers, for example *Harpagoxenus americanus*, and these forms have a very small behavioural repertoire compared with less specialized slave-makers such as *Leptothorax duloticus* (Wilson, 1975a). A similar comparison could be made between the highly specialized *Polyergus* and *F. sanguinea*. *H. americanus* and *Polyergus* workers are so specialized that they can hardly survive when deprived of their slaves—again this is often regarded as a symptom of their degeneracy, but more correctly represents the trend seen in the most advanced polymorphic societies. Certain *Pheidole* majors, for example, can barely survive if deprived of their minor working sisters (Wilson, 1984).

For this reason a modern evolutionary evaluation of the adaptations of ant parasites might best proceed by using the analogy that these exploiters represent an interspecific experiment in life-history strategies, and the design of a division of labour.

CHAPTER EIGHT

ANT ECOLOGY

Ants occur in almost every terrestrial habitat, from tundra north of the Arctic Circle (Brown, 1955) to the tip of Tierra del Fuego (Wilson, 1971). In many ecological communities ants are so dominant that all the available space appears to be the province of one colony or another, and they occupy a large number of roles in each community. They are important as predators, scavengers and consumers of seeds and in a wide variety of mutualistic interactions with many families of insects and plants (see Chapter 6). In many deserts ants are major consumers of the seeds of annual plants and actually depress the population growth rate of rodents who depend on the same food (Brown and Davidson, 1977). By contrast, in many temperate northern woods ants act as unwitting gardeners dispersing the seeds of approximately 30% of the herbaceous flora (Beattie, 1985). Ants may even be important in the formation of certain soils (Lyford, 1963). Though certain species have specialized, for example as seed harvesters, as farmers of Homoptera, or as leaf-cutters growing fungus gardens, most of the 12 000 or so living ant species have a mixed diet which allows for great plasticity in their foraging ecology. In Amazonian forests, ants occupy a greater diversity of roles than perhaps anywhere else and, together with termites, constitute 30% of the total animal biomass (Fittkau and Klinge, 1973).

Rather than examine in detail this great diversity in the foraging and nesting behaviour of ants which affect the growth of ant populations in innumerable ways, this chapter will explore the two almost universal features of ant ecology. First, within many habitats the most important factor that determines the distribution and abundance of ants is the distribution and abundance of other ants. Second, the ability of ants to dominate space is often severely limited by the necessity of foraging and patrolling from a fixed nest site. The crux of the problem of this so-called central place foraging (Brown and Orians, 1970) is that as a colony grows, foragers must travel further and further at ever-increasing cost to retrieve less and less energy per unit time. That is, more and more resources must be used merely to provide a larger foraging population and to pay their travel costs.

We will consider in turn how ants compete for limiting resources, how they defend such resources, what rules they might use to forage efficiently and which prey they should select to maximize their rate of gain of energy and other resources. Lastly, since competition among ants might be expected to reduce their species diversity through competitive exclusion, we will review some of the factors that maintain diversity in ant communities.

8.1 Competition

Competition among ant colonies, which limits the productivity and fitness of colonies, can occur in two different ways: either colonies directly contest or scramble for limiting resources, or they have a directly detrimental effect on one another through other forms of mutual interference. For most ant colonies the intensity of the exploitation or interference competition they suffer will largely depend on the species, size and proximity of their neighbouring colonies. This is simply a consequence of the nests of most ants being almost permanently rooted in place. For this reason, Brian (1965) has drawn attention to the similarity between the ecology of ants and of plants. The success of an individual plant or of a single ant colony depends to a great extent on the abundance of resources in its immediate neighbourhood and how these resources are limited or controlled by its neighbours.

For ants and other sedentary populations space can be considered a primary limiting resource. Support for this assertion comes from a wide variety of different ecological communities, in which perhaps the single most common finding in studies of the distribution and abundance of ants is that their colonies tend to be spaced out in a regular way (Levings and Traniello, 1981). This immediately suggests that where a colony can live and grow depends on the position of other colonies.

Patterns of nest spacing can be investigated by a wide variety of different tests (Southwood, 1978). One of the most useful for ants involves a comparison of observed distances between nearest neighbours with expected nearest-neighbour distances predicted on the basis of a random (i.e. Poisson) distribution (Clark and Evans, 1954). If the position of a nest is independent of the positions of other nests then spacing is predicted by a

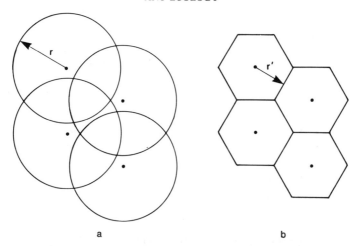

Figure 8.1 (*a*) When ants forage for evenly dispersed resources they are likely to search in a circular area centred on the nest. However, when the potentially circular foraging areas of neighbouring colonies overlap and there is defence of territorial boundaries, the circular areas tend to collapse into the shape of polygons such as the hexagons drawn (*b*). If such territorial behaviour is further associated with the destruction of new colonies by established ones, the population of nests will tend to stabilize in a regularly dispersed pattern.

Poisson distribution. However, if each nest inhibits the occurrence of others near it, then nests would tend to be more equally spaced than a Poisson distribution would predict. Perfect maximum spacing would occur if each colony occupied the centre of its own hexagonal space; the entire area would then be divided up into a lattice of the largest possible non-overlapping hexagons with one for every colony in the population (Poole, 1974) (Figure 8.1).

Levings and Traniello (1981) have made a thorough review of the literature on intra-specific spatial patterns in ant communities. In 80 reported case studies that could be examined quantitatively, 67 showed a significant tendency towards regular spacing. In another 80 cases that could not be treated statistically the majority of reports strongly suggested an even dispersion pattern. Regular spacing of ant colonies is seen, for example, in English grasslands, heathlands and woodlands, in Australian and New World deserts and savannahs, and in tropical forest in both hemispheres. The species showing such regular spatial patterns include specialist predators, scavengers, aphid farmers, seed-eaters, leaf-cutters and generalist omnivores. Thus there is a marked trend towards regular spacing in most ant species in which nesting patterns have been examined, however much habits and habitats differ from species to species. In addition a regular spacing of nests can be detected not only within species but also between different ant species (Elmes, 1974; Bernstein and Gobbel, 1979; Levings and Franks, 1982) suggesting that interspecific competition also occurs for foraging and nesting space.

A regular spacing pattern of nests is important evidence for competition, for the simple reason that it is unexpected unless competition is extremely intense. Both the patchiness of resources and widely different sizes of ant colonies (and hence differences in the amount of space they can dominate) would be expected to override any tendency for ant nests to be regularly spaced. For the same reasons regular spacing is even less expected among colonies of different species unless interspecific competition is particularly strong. The interplay of these various factors has been shown in two studies in very different habitats.

Bernstein and Gobbel (1979) documented spatial patterns in ant communities in the Mojave and Great Basin deserts of North America. Their study is worth considering in some detail here because many of the ants in this community will be referred to again in later sections of this chapter on optimum foraging and interference behaviour. The species they recorded included the dolichoderine omnivores, *Iridomyrmex pruinsosum* and *Conomyrma bicolor*, formicine omnivores such as honey-pot ants of the genus *Myrmecocystus*, and a number of myrmicine seed-harvesting ants including two species of *Pheidole*, two forms of *Veromessor* and four species of *Pogonomyrmex*. Among these ants regular spacing of nests could be found both within species and at the level of the entire community. However, regular patterns occurred only in deserts at low altitudes. All nests considered together were regularly spaced at lower elevations, randomly spaced at intermediate elevations and clumped at higher altitudes. Bernstein and Gobbel (1979) suggest that the tendency for regular spacing in the ant community as a whole is overridden at higher altitudes because unshaded areas, which are needed for the establishment of nests, are clumped between the perennial plants that are more common at higher elevations.

Levings and Franks (1982) conducted a similar study in a completely different habitat. They examined nest spacing patterns in the ant community of the floor of the rainforest on Barro Colorado Island, Panama. This island supports a spectacularly rich ant fauna: well over 100 species of ant forage on the ground layer of the lowland tropical forest and perhaps twice that number occur in its canopy. Levings and Franks (1982) mapped a selected group of 15 species of the most common omnivorous ants. All 15 species were either ponerines or myrmicines, that forage during the daytime. The former subfamily was represented by species such as *Pachycondyla*, *Ectatomma* and *Odontomachus* with large workers who forage alone, whereas the myrmicines were mostly species of *Pheidole* and *Solenopsis* with tiny workers whose activity is coordinated by pheromone trails which they use to recruit large numbers of workers to large prey items. In all mapped areas the arrangement of all nests considered together was significantly regular, and in the majority of cases regular spacing also occurred at both the subfamily and the species level. This was all the more remarkable because nests of the two subfamilies were consistently randomly intermingled, so that the overall regular spacing could not be attributed merely to competition within a particularly abundant species or subfamily. The nearest neighbour of a ponerine colony was no more likely to be another ponerine nest than a myrmicine nest or vice versa, yet each type of colony was regularly spaced from its neighbours no matter who they were. All this evidence suggests that, in spite of their very different foraging methods, these colonies treat one another as ecological equivalents with regard to spacing. This is most surprising because the myrmicine workers of the species in this study each had a dry weight that was less than one-twentieth that of the individual ponerine workers. *Pachycondyla* workers can individually monopolize and retrieve large prey items that would require the efforts of a group of *Pheidole* workers. Yet notwithstanding this fundamental difference in their foraging behaviour, which must play a major part in deciding which type of established colony would win in a direct battle over their food supply, such colonies appear to be equally efficient in denying space to one another.

This finding can be explained most easily not by considering competition among established colonies but in terms of the attacks by established colonies on new ones. Indeed, the very same aggressive interference may be the chief factor involved in all the above cases of regular spacing of ants' nests. This is a special and very powerful form of interference competition. The destruction of new nests by foragers from established colonies has been seen in a wide variety of species in a large number of different habitats (Pontin, 1960, 1963; Brian, 1965; Carroll and Janzen, 1973) including deserts such as those studied by Bernstein and Gobbel (1979) (see for example Hölldobler, 1976*b*; Hölldobler *et al.*, 1978).

It might be thought that the destruction of new nests is purely the result of most ants being at least in part predators of other insects. For such ants foreign queens would represent a valuable source of protein. However, even such obligate consumers of plant material as harvester ants and leaf-cutters are also known to destroy colonies of the same and other species (Hölldobler, 1976b; Jutsum, 1979). Thus it is probably safe to conclude that the destruction of potential neighbours is a distinct competitive strategy whose advantage is the protection of foraging space from potentially avaricious neighbours rather than merely the incidental intake of extra insect protein. Defence of an area by eliminating a small colony before it possess a threat to limited resources will cost less while the established colony is much larger than the new one. Such defence will also be more cost-effective for long-lived colonies whose nests occupy a fixed position. Furthermore, such pre-emptive exclusion can be more easily extended to many species of potential competitor who might forage in different ways or at different times of day and therefore would be difficult to exclude by other forms of territorial defence. Its benefits, however, will be felt over a much longer time-scale as it prevents a day-to-day running battle with ever-growing neighbours. In this way, the processes that determine which established colonies survive to compete with each other in normal ways may have occurred years or even decades before long-lived colonies grew to their present size.

Since attacks on small colonies by large ones are very likely to succeed, much more attention should be paid by ecologists to the strategies queens of new colonies use to escape detection by other ants. The three major forms of colony foundation in ants—claustral, primary pleometrosis, and fission—are described in different contexts in Chapters 1 and 3. Each, in its own way, can be a strategy to maximize the survival of young queens and new colonies in situations in which there is intense competition from established colonies.

Evidently colonies have the ability to prevent the formation of small new colonies of almost any species that settle too close; however, the worst type of neighbour is still one of the same species. Conspecific neighbours pose the greatest competitive threat because they will forage in the same way, at the same time of day and for the same resources. Their elimination would bring the greatest benefit, and for this reason selection should cause foragers to destroy colony-founding queens of their own species preferentially. In addition such queens may simply be easier to detect. Just such a preference has been demonstrated by Pontin (1960) who showed that *Lasius niger* and *L. flavus* workers are more likely to attack conspecific queens than those of the other species. Greater tolerance for other species than for conspecifics is probably one of the most important factors that act to maintain diversity in ant communities (Levings and Franks, 1982).

Janzen (1970) suggested a mechanism that could maintain the diversity of tropical rainforest trees and could equally well apply to ant communities. According to Janzen's hypothesis seedlings near to an established plant are exposed to large numbers of species-specific herbivores and seed predators. Hence the only seedlings to survive are those at some distance from an established conspecific. In this way each host-specific guild of herbivores creates space in which other tree species can become established. This is an important hypothesis because the search for an explanation of how hundreds of tree species can coexist in the tropics has been called one of the great problems in modern ecology. Explaining the diversity of ants, particularly in the tropics, is no less of a problem. Established ant colonies directly destroy incipient colonies of the same species preferentially, just as an established plant indirectly by its herbivores may destroy conspecific seedlings. In both plant and ant communities the forces that maintain pattern diversity may also maintain species diversity by creating space in which other species can survive.

The importance of space as a primary limiting resource in both intra- and interspecific competition in ants and its effect on the fitness of colonies has been clearly shown in the studies of Pontin (1961, 1963). He examined in detail the ecology of two common British ants, *Lasius flavus* and *L. niger*. Both species build large nest mounds in English meadows and grasslands. *L. flavus* predominantly forages underground whilst *L. niger* mostly forages above the soil surface. However, removal experiments suggest that one-third of the food of *L. niger* colonies comes from below ground to the detriment of *L. flavus* (Pontin, 1963). Notwithstanding the substantial difference in the foraging ecology and diet of these species, competition between them is intense and takes the form both of predation of new queens by old colonies and of direct competition between established nests. However, in either case intraspecific competition is stronger than interspecific competition and this is an important factor in the stable coexistence of these species. Each species inhibits its own growth more than that of the other, so that both species can coexist.

The strength of competition between established colonies has been demonstrated by experiments in which colonies were excavated and replanted in different sites. Not only do established colonies kill unrelated queens, but their own production of sexuals can be shown to be limited by the proximity of competing colonies of both the same and the other species. This is the clearest evidence that spacing affects fitness. Pontin (1963) showed that the number of queens (n_1) and (n_2) produced by two neighbouring colonies of *L. flavus* is positively correlated with their distance (D) apart. A line of best fit through the scatter plot of $(\sqrt{n_1} + \sqrt{n_2})$ against D is straight

Figure 8.2 Spacing affects the production of new queens by colonies of *Lasius flavus*. The graphs show the relationships between the distance (D) between pairs of colonies and the number of queens N_1 and N_2 each produced, in 1956 and 1957. Such relationships are expected if colonies have feeding territories that are limited by the proximity of neighbours. (Redrawn from Pontin, 1963.)

and crosses the axes close to the origin (Figure 8.2). Just such a relationship is expected if colonies have circular territories, whose area determines the amount of food they gather and hence the number of sexuals they can produce. Pontin (1961) also showed that queen production by *L. flavus* could be reduced to one-quarter of its previous level by the presence of adjacent *L. niger* colonies. Recently, Boomsma et al. (1982) have confirmed these findings and have shown that the production of queens by *L. niger* is also reduced by 74% by neighbouring *L. flavus* colonies; a remarkable similarity. In addition, in sand dune habitats in Holland, not only the number but also the sex ratio of winged offspring of *Lasius* colonies is affected by competition (Boomsma et al., 1982). Thus in previously unforeseen ways, territory size and spacing affect the inclusive fitness of ant colonies. This brings us to the problem of territory design. What is the optimum size and shape of the area that a colony should defend to maximize its fitness?

8.2 Economics of territorial defence

If territorial behaviour is subject to evolution by natural selection, animals should defend the appropriate size and type of territory to provide themselves with the most benefits for the least costs. Benefit may be equated with energy for growth and reproduction, and costs include the energy invested in defence and the risk of death or injury associated with driving off competitors (Davies, 1978; Brown, 1964). A territory can be a key aspect of a colony's infrastructure as described in Chapter 3. Davies and Houston (1984) provide a recent review of the economics of territoriality, drawing their examples mostly from studies of vertebrates. Ant territoriality is rather different for two reasons: first, battles can be fought to the death by altruistic workers of eusocial insects. The short-term loss to the colony can be compensated by a long-term gain if resources accrue from such conflict. Solitary or semi-social vertebrates, on the other hand, must defend their own territory without incurring injuries which would impair their reproduction. Second, eusocial ants can approximate to the ideal defence of being everywhere at once: the absolute territory is within reach of the insect society but eludes the lone vertebrate who can only guard one part of its domain at any one time.

Hölldobler and Lumsden (1980) have shown that for ants the optimum territorial strategy depends not only on the distribution of resources within the habitat but also on the social design of the colony infrastructure, including the dispersion of the nest or nests, their recruitment techniques,

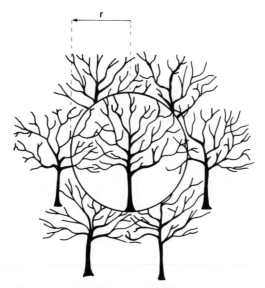

Figure 8.3 Economics of territorial defence. A model for the weaver ant *Oecophylla* whose territories can be 40 m across and span the full depth of the canopy. The cost of defence is proportional to the surface area of the boundary of the territory, and the benefit is proportional to the volume of the space, and hence the amount of food, defended. r is the radius of the hypothetical spherical territory for *Oecophylla*. (Redrawn from Hölldobler and Lumsden, 1980.)

chemical signposts of territorial boundaries, and even techniques for the physical intimidation of competitors. Hölldobler and Lumsden (1982) considered as an example the weaver ant *Oecophylla* which defends huge three-dimensional territories (Figure 8.3). Throughout much of Africa these ants are rulers of the tree-tops. They can defend their exclusive territories against almost any enemy, including driver ants of the genus *Dorylus* (Gotwald, 1982).

One special feature of *Oecophylla* is that their colonies' nests and food resources are both dispersed evenly throughout the territory. Colonies prey on or collect honeydew from insects that are evenly scattered over the vegetation of the canopy as are the nests of the ants themselves. A highly dispersed nest is possible for this species because *O. longinoda* workers can make nests almost anywhere in the trees, near their food sources, by forming leaves into envelopes that are woven together at the edge with silk from the larvae (see Chapter 3). Colonies of these ants have a single queen and up to 500 000 workers.

Hölldobler and Lumsden (1980) present a model that illustrates the factors determining territory size and shape for *Oecophylla* colonies. The costs of defending a three-dimensional territory will depend largely on its external surface area, as this is the boundary that must be patrolled, whereas the benefit from such defence is likely to be a function of its volume. If the territory is spherical the volume/surface ratio rapidly increases with

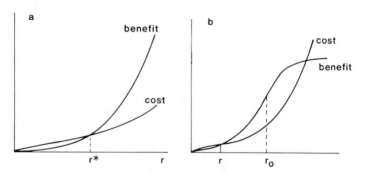

Figure 8.4 Graphical models for territorial defence by *Oecophylla*. Cost and benefit curves for a boundary defence strategy on spherical territories. The x axis represents territory radius (r). (a) The symbol r^* represents the threshold below which any defence is uneconomic. The benefit and cost curves diverge because large spheres have larger volume to surface ratios (b). However, more realistically, very large territories would have increasing travel costs associated with carrying eggs or larvae from the central queen-nest to the peripheral nests. Thus toward large r the benefit curve tips over. Maximum benefit to cost is achieved for territories of radius r_o. (Redrawn from Hölldobler and Lumsden, 1980.)

size (Figure 8.4) and therefore a very large territory would be most profitable. However, for most ants a large territory would be unmanageable as they would waste too much energy moving prey items from the periphery to the central nest. *Oecophylla* has alleviated this problem by moving its brood to its food—by building leaf nests all over the territory. Clearly a colony still has to move its larvae or eggs from the queen in the centre of the territory to the peripheral nests. However, fewer journeys will be necessary to do this than to return repeatedly with food to a central crèche. As the theory suggests, *Oecophylla* colonies have enormous territories which may be 40 m across and take up the entire depth of the canopy. The cost of territorial defence is also reduced through their use of a true territorial pheromone. This serves as a 'trespassers will be prosecuted' sign for foreign ants. When workers enter new terrain they deposit fluid from their rectal sack which will induce aversive behaviour if it is encountered later by alien conspecifics. The major limitation on territory size in *Oecophylla* is probably colony size, which must be limited in part by the egg-laying rate of the single queen.

By contrast, harvester ants of the genus *Pogonomyrmex* have to face problems associated with central place foraging. They have a single nest, and their food resources—in this case seeds—are distributed in small discrete patches, whose only saving grace is that they are persistent. To model the economics of territorial defence in this species, Hölldobler and Lumsden (1980) envisioned the area around the nest as a series of sectors of small and equal size, like sections of a pie diagram (Figure 8.5). Only some of these sectors will contain food. The value of a particular food patch to a colony will depend on the size of the patch, its distance from the nest and also whether it has been discovered by a foreign colony. Simple models predict that sectors that are known to be empty or to contain only small, distant or already defended patches should not be patrolled at all. The most economic solution is a trunk trail system in which foragers are channelled towards known patches. This is exactly what occurs in *Pogonomyrmex barbatus* and *P. rugosus* (Figure 8.6). New resource patches, which do not belong to competitors, are rapidly explored and occupied by workers who have made individual excursions from the ends of existing trunk trails, and extra workers are quickly brought to the new site through the use of chemical recruitment systems. The speed of recruitment depends on the characteristics of the food source: its distance from the nest, the density of seed fall, and the size of the grains as well as the presence or absence of foreign foragers at the resource patch (Hölldobler, 1976b). Seed sites previously occupied by competing foragers were considerably less attrac-

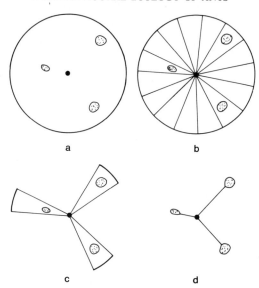

Figure 8.5 A diagrammatic representation of possible *Pogonomyrmex* (harvester ant) territories. Small discrete patches of harvestable seeds are scattered around the central nest. The ants might patrol and forage in discrete sectors. Only seed-containing sectors should be patrolled. This economizing strategy finally results in trunk trails leading to known food patches which are the sole part of the foraging area that is defended against neighbouring colonies. Such a scheme results in the pattern of trunk trails shown in Figure 8.6. (Redrawn from Hölldobler and Lumsden, 1980.)

tive than unoccupied seed sites. Hölldobler and Lumsden (1980) concluded that the trunk trails of *Pogonomyrmex* are partitioning devices which curtail aggressive confrontations between neighbouring colonies, while at the same time maximizing the number of food patches in a given area foraged. They also noted that similar foraging and space-partitioning systems occur in several species of *Formica*, and in leaf-cutter ants of the genus *Atta* and in *Lasius neoniger* which uses a colony-specific trail pheromone to mark its trunk routes (Traniello, 1980). In all these cases resources occur in small, discrete and relatively persistent patches.

The honey-pot ant, *Myrmecocystus mimicus*, faces an even more complicated problem of resource defence than either *Pogonomyrmex* or *Oecophylla*. *Myrmecocystus* has the dual problem of central place foraging and resources that are unpredictable in both space and time. Honey-pot ants occur in the southwestern deserts of the United States and eat a wide variety of insects; however, where and when a patch of these prey will be

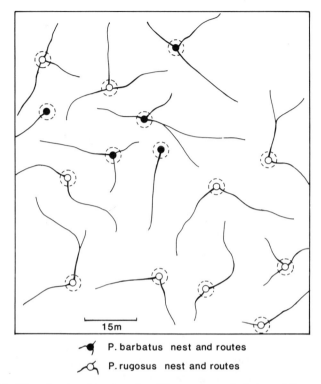

◀︎ P. barbatus nest and routes

⚆ P. rugosus nest and routes

Figure 8.6 Map of nests and trunk routes of *Pogonomyrmex barbatus* and *P. rugosus* in New Mexico. Solid black circles *P. barbatus*: open circles *P. rugosus*. (Redrawn from Hölldobler and Lumsden, 1982.)

discovered cannot be predicted. To model the problem faced by a honey-pot ant colony, first consider a purely hypothetical case. Imagine that colonies hold open, exclusive, circular territories, and that the cost of defence is proportional to the area of the territory. If patches occur at random in space and time then the probability that a colony will encounter zero, or one or more, resource packets in a certain time in its territory, can be modelled using the Poisson distribution. (This statistical distribution can be used, for example, to predict the number of randomly thrown packets that will drop on average into a series of boxes.) The probability that a colony holding open a territory of area a will receive one or more resource packets in a period of time T is given by

$$P = 1 - \exp(-aD)$$

where D is the average density of resource packets occurring per unit area in a time interval T. As a and/or D increases then the probability of encountering resources will also increase. In a sense, by holding open a large territory a colony would be gambling that more resources will drop into the defended area, at a rate more than sufficient to compensate for the cost of defence. This process is similar to purchasing more lottery tickets in the hope that one will be a winner and more than cover the increased total cost of entering the lottery. But does it pay to gamble? The Poisson distribution tells us that if packets are rare and the boxes small, then at any one time the majority of boxes will be empty. However, even if resources are more common but not abundant, or the area searched is very large, then the hypothetical colony will have extremely unpredictable returns for its efforts. The mean number of packets n occurring in a time interval T in an area a will be given by $a \times D$. But the Poisson distribution tells us that the variance around this mean is equal to the mean itself. Thus in situations in which the success rates of colonies can be described by the Poisson distribution and other similar statistical distributions it is clear that they will suffer constant costs of defence for extremely variable, and often small, benefits. It is unlikely that the best strategy for a colony in such an unpredictable situation would be to engage in such a risky gambling game, even if it could store food from winning streaks to carry it over the bad times.

Myrmecocystus does have the ability to store food, because it has a special replete caste of workers who become living storage pots. The gasters of such workers swell to a huge size as they become distended by the crop which is filled with the juices derived from their food (see Chapter 2). Nevertheless, exclusive territories are likely to yield exceedingly unpredictable gains even for these honey-pot ants. Hölldobler and Lumsden (1980) suggest as a general rule of thumb that animals that would be faced with chains of deficits and surpluses on territories of fixed size should dissolve their boundaries and let foraging ranges overlap. Such animals should adopt the strategy of spending their energy on reconnaissance followed up with rapid, exclusive exploitation once resources have been discovered. This is exactly what occurs in *Myrmecocystus*, with an added twist: colonies physically intimidate rivals by engaging them in elaborate tournaments if they find a resource patch that might also be discovered by the rival colony. Thus honey-pot ants expend energy on discovering food over a large area rather than defending the whole area, and then they defend food sites only when they are known to be profitable. Because such sites vary in space and time, such behaviour is best described as spatio-temporal territoriality.

A major source of food for *Myrmecocystus* is termites which forage on dispersed dead vegetation. When a scout discovers a rich patch of termites it brings a group of its nestmates to the site by means of special recruitment signals. If a second colony of *Myrmecocystus* is discovered close to the site, some foragers from the first colony return home and summon a large group of their nestmates to the entrance of the foreign colony. The result is an extraordinary tournament: workers from the first nest engage all the workers emerging from the second in a ritualized display, in which the individual workers draw themselves up to their greatest height by walking on stilt-like legs as they parade past members of the opposite camp (Figure 8.7). Each group of displaying ants breaks up after 10–30 s, but the ants retain their elevated posture. When they meet a nestmate, they respond with a brief jerking display, but when they meet another opponent the whole aggressive display ceremony is repeated (Hölldobler, 1981). This prevents foragers leaving the second colony and getting to the prey. The tournaments probably represent an exchange of information on the relative size of the opposing colonies. There are two possible sources of information on colony size: (i) the number of workers that a colony can summon to a display and (ii) the size of the individual workers. Bigger colonies usually have larger workers (Chapters 3 and 4): hence the ants walking on extended legs to appear as large as possible. When one colony is very much larger than the other the tournament may end in an intraspecific slave raid (Chapter 7) with the smaller nest being destroyed. In other words, in *Myrmecocystus*, sophisticated interference behaviour can revert to the almost ubiquitous form of territory defence of intraspecific destruction of a new nest by an established one. Resources can be defended by complex

Figure 8.7 Two honeypot ants, *Myrmecocystus mimicus*, from different colonies engage in a ritualized stilt-walking tournament as part of the defence of a spatio-temporal territory by one of the colonies. See text for further explanation. (Drawn from a photograph by B. Hölldobler.)

interference behaviour between colonies of different species as well as within species. Two particularly remarkable examples also occur in the deserts of the southwestern United States. In both cases *Myrmecocystus* suffers interference from colonies of other species. Workers of the dolichoderine ant *Conomyrma bicolor* drop stones down the entrances of the *Myrmecocystus* nests (Möglich and Alpert, 1979). This behaviour apparently prevents the honey-pot ants foraging. Another dolichoderine ant, *Iridomyrmex pruinosum*, chemically interferes with *Myrmecocystus* workers at the food source as well as at the nest entrance, to reduce the foraging activity of its competitor (Hölldobler, 1982).

As we have seen, it probably would not be profitable for *Myrmecocystus*, *Conomyrma* or *Iridomyrmex* colonies to defend the whole of a designated area around their nest because food resources crop up too rarely. The interference behaviour of all three species is only likely to be called into action if they discover resource patches that cannot be quickly retrieved by a single worker. Furthermore, such interference behaviour relies on sophisticated recruitment techniques and rapid deployment pheromones. From this reasoning it is clear that ant colonies that forage for small, relatively rare and randomly occurring prey that can be retrieved by individual workers, should have no other territorial defence than the destruction of small neighbouring nests.

8.3 Foraging for the most profitable prey

Just as natural selection should favour colonies who control the optimim size of territory, evolutionary pressures should also lead to individual workers choosing the most profitable prey. In this field too optimization theory can provide useful guidelines for investigating the decision-making and performance of ants.

Foraging workers frequently face the problem of whether they should retrieve a certain type of prey or continue to search for a more profitable one. The appropriate solution to this dilemma, if prey differ only in their energy value and not in essential nutrients, would be the one that helps the colony maximize its net gain of energy per unit time. To simplify the analysis it is necessary to assume that all prey can be ranked by workers in discrete categories, from best to worst, according to their expected energy yield. For central place foragers, such as ants, Orians and Pearson (1977) have proposed that prey might be ranked by estimates of their profitability E/h, where E is the expected net energy content (i.e. benefit) of a food type and h represents the time an item would consume in terms of pursuit, capture, handling and retrieval (i.e. its cost). All the remaining time t available to a forager is assumed to be spent searching for a new prey item.

To maximize their rate of energy gain, ants should always take the highest-ranking prey in terms of these criteria of benefit and cost. A more interesting question is what other categories of prey the ants should include in their diet. As more types of prey are included, prey is encountered more often and the search time t will decrease, but so will the average profitability of captured prey. Hence there will be some optimum trade-off point between the opposing advantages and disadvantages of including more items in the diet. This point can be visualized graphically (see Figure 8.8).

In addition to taking the best prey whenever it is encountered, the ants should add other prey types to their diet in order of decreasing rank of profitability, only if they are sufficiently profitable to increase the overall energy gain per unit time. This argument is equally applicable to different categories of prey types and to different categories of prey patches. For example, when should workers visit patches of the same quality and density of prey at different distances from the nest? Such optimization arguments

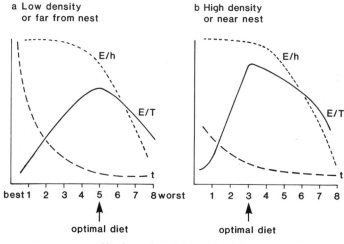

Figure 8.8 A graphical model of optimal breadth of diet. As more prey are addded, travel time (t) decreases, but so does the average profitability of prey consumed (E/h); the curve of net food intake per unit time ($E/T = E/t + h$) rises to a peak and then declines. The peak is the optimal diet breadth. (*a*) When prey are at low density or far from the nest, the ants should be less selective than when (*b*) prey are at high density or near the nest. The highest-ranking, most profitable prey should always be retrieved wherever they are encountered. (Redrawn from Krebs, 1978.)

are by necessity highly simplified: for example we assume that every ant has a complete knowledge of its foraging environment and of the abundance and value of all prey and foraging sites. This is highly unrealistic, but nevertheless these models still provide valuable tools for investigating the foraging ecology of ants.

In field studies with ants in North American deserts, Davidson (1978a) investigated the foraging selectivity of two species of harvester ants, *Pogonomyrmex rugosus* and *P. barbatus*. To eliminate differences in the nutritional quality of the seeds they forage for, she provided colonies with non-native pearl-barley seeds in four different sizes. In these field experiments native seeds were at such low densities that they were completely ignored when pearl-barley seeds were introduced. In the first set of experiments, *P. barbatus* colonies were alternately exposed to high and low densities of the barley seeds with the relative abundance of the four seed sizes held constant. The seeds were arranged in circular patches 3 m from each colony's entrance. At low densities more time is consumed in searching, so the value of any item, large or small, is relatively greater than it would be at higher densities. For this reason dietary breadth is expected to be greater at lower prey densities. This is exactly what was found in these experiments. At low seed densities the ants took significantly more small seeds.

In the second set of experiments *P. rugosus* colonies were presented with a seed patch at 3 m, 6 m, 9 m or 12 m from the nest. Again each patch consisted of four size-classes of barley particles in equal proportions by weight. At all distances, the ants preferred seeds from the next-to-largest category because the largest seeds were hard to handle. In addition, ants took an increasingly large proportion of seeds from this preferred size class the greater their distance from the nest. Again this fits the prediction of the model. Simply, smaller seeds have even less value at greater distances from the nest.

Close observation suggested that individual workers handled a number of seeds before selecting one to carry back to the nest. In this way individuals probably reach their own independent decisions about which items to choose. In Davidson's experiments *Pogonomyrmex* workers chose seeds on the basis of size. In the experiments of Taylor (1977) both *P. occidentalis* workers foraging for seeds and *Solenopsis geminata* workers taking liquid food were able to partition their efforts between food patches in ways that followed the very precise predictions of optimality models that analysed energetic efficiency directly.

Efficient foraging also depends on the competitive environment of a colony. Davidson (1978b) has also shown that for another harvester ant, *Veromessor pergandei*, seed-size preferences are related to the size of the foraging workers. Where many species compete for the same seeds, *V. pergandei* workers have a narrow range of sizes and hence a narrow dietary breadth. Where competition is less intense and a wider range of seeds is available, *V. pergandei* workers are more polymorphic and have a wider dietary breadth.

8.4 Ants as predators and prey: army ant foraging ecology

The major themes of this chapter, competition for foraging space, interference behaviour, foraging efficiency and the dynamics of ant communities, are all important in the foraging ecology of army ants.

Among all ants, army ants have the most advanced systems of group foraging. By raiding in very large numbers they are able to overcome larger prey than are available to solitary foragers. All army ants prey upon such organisms as large arthropods, social insect colonies or even earthworms (Gotwald, 1982). These prey can only be tackled by large numbers of foragers, raiding from large colonies. Such raiding denudes large areas of these prey which are slow to recover, and so all army ant colonies show elaborate and frequent emigration behaviour.

Army ants are a polyphyletic group and occur in North and South America, Africa, Asia and Indo-Australia. There are more than 140 species of army ant in the New World alone (Watkins, 1978). One of the most thoroughly studied species, *Neivamyrmex nigrescens* lives in the deserts of North America, whose ant communities have already been referred to many times in this chapter, These *Neivamyrmex* colonies show a decided preference for raiding *Pheidole* nests (Mirenda *et al.*, 1980).

However, of all army ants the ecology and behaviour of *Eciton burchelli* colonies is perhaps the best known. This species inhabits lowland tropical rainforest from Peru to Mexico, but has been most intensively studied on Barro Colorado Island, in the Republic of Panama. *Eciton burchelli* colonies stage the largest raids of any neotropical army ant: a single swarm raid may contain up to 200 000 ants and can be 20 m wide (Willis, 1967; Franks, 1982c). The raid moves constantly forward as a phalanx of ferocious workers, and such is the density of ants in the swarm that they make the floor of the rainforest look like a seething brown river of marauding insects. During a day's raid the swarm front proceeds in a zigzag pattern; a turn to the left is followed by a turn to the right and vice versa, so the overall course of a raid is roughly a straight line (Figure 8.9). The

180   THE BEHAVIOURAL ECOLOGY OF ANTS

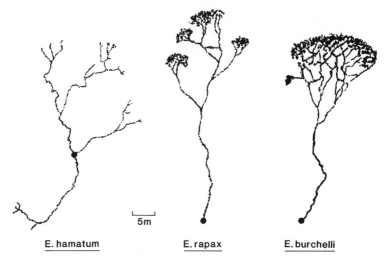

Figure 8.9 A comparison of the foraging patterns of army ants. *Eciton rapax* has an raid system intermediate to that of the column raiding of *E. hamatum* and the swarm raiding of *E. burchelli*. The large circles indicate the positions of the bivouacs. In all cases a principal raid column links the foraging ants to the bivouac nest. (Redrawn from Burton and Franks, 1986. The foraging patterns of *E. burchelli* and *E. hamatum* are redrawn from Rettenmeyer, 1963.)

swarm moves at an average rate of 14 m per hour, which is so slow that almost any terrestrial vertebrate can outrun these predators. However, the raiding army ants capture a wide variety of large arthropods such as roaches, spiders and scorpions, though most prey are ant species that nest on the forest floor. An experiment in which an entire colony of these army ants was taken to Orchid Island, a neighbour of Barro Colorado Island that was previously free of *Eciton*, showed that even these most polyphagous army ants are would-be specialist predators of other ants (Franks and Bossert, 1983). On Barro Colorado Island, 55% of the prey of *E. burchelli* colonies is derived from the nests of other ants, but on Orchid Island 86% of the introduced army ant colony's prey was ants, and this difference was maintained through both wet and dry seasons (Franks and Bossert, 1983). Before the introduction, the ant prey of *E. burchelli*, mostly colonies of *Camponotus*, *Pachycondyla* and *Odontomachus* were more than twice as common on Orchid Island as on Barro Colorado Island. Furthermore, spatial pattern analyses showed that these abundant prey colonies, in the absence of their major predator, were displacing other ant species from the leaf litter community. As Leving and Franks (1982) showed, many of the ants of this leaf litter community treat each other, at least in terms of competition for nesting space, as ecological equivalents. This raised the possibility that by removing or reducing the size of certain ant colonies on the forest floor, *E. burchelli* could create space in which other ant species flourish. Areas recently raided by *E. burchelli* were compared with other areas of the forest floor by using baited pit-fall traps to capture workers of the ant species that live and forage in the leaf litter. Transects of these pit-fall traps were sampled at intervals over several months.

A raid of *E. burchelli* initiates a process of change in the ant community which resembles the well-known process of succession in plant communities. The first stage in this succession in the aftermath of a raid is characterized by the rapid increase in the abundance of one species of *Paratrechina*. Members of this formicine genus are classic opportunists. They occupy new areas not just through the formation of new colonies but also through the growth of existing ones; colony emigrations are frequent and typically they invest only in flimsy unstable nest sites that would be unsuitable for other ant species. These tramp-like habits are facilitated by their unicolonial social organization with no aggressive interactions or boundaries between colonies (Hölldobler and Wilson, 1977). In the aftermath of swarm raids there are more species of ants of the genus *Pheidole* than in other areas. Furthermore there is a greater turnover of these species in raided areas than in areas not recently foraged. The rapid influx of *Paratrechina* may in part cause the rapid turnover of *Pheidole* species in raided areas, but a turnover of these *Pheidole* species could also have been predicted by the work of Levings and Franks (1982) who showed that *Pachycondyla* and *Odontomachus* species deny space to these myrmicines. Thus when these ponerines are cropped by an army ant raid, new *Pheidole* colonies can become established in the leaf litter.

Army ant raids also favour the establishment of colonies of their prey species. The results of an experiment with artificial nest sites demonstrate an interaction between the predation of queens by established colonies and predation of established colonies by army ants. Incipient colonies of prey species are founded in greater abundance in artificial nest sites placed in recently raided areas than in similar nests in control areas (Franks, 1982*b*). The simplest explanation is that in the aftermath of swarms, when prey colonies have been killed or cropped to smaller size, intraspecific predation of queens of prey species is reduced, with the result that new nests of these species can be initiated more successfully. Baited pit-fall traps were also used to estimate the regrowth of the ant populations within raided areas. These data suggest that the ant fauna takes about 200 days to recover fully following an army ant raid.

But how important are army ant raids in the overall dynamics of the leaf litter community? How often do army ants raid areas; and what fraction of the forest floor is recovering from a raid at any one time? These questions can only be answered by a detailed examination of the search pattern of individual *E. burchelli* colonies. This is simpler than might be expected because these army ant colonies have a rigid temporal pattern of activity, associated with the development of their brood, that to a large extent dictates their foraging pattern. *E. burchelli* colonies maintain, throughout their lives, a 35-day cycle of activity (in which they alternate between 20-day statary phases in which they maintain the same nest site and raid on only 13 days on average) and 15-day nomadic periods (in which they raid every day and emigrate to a new bivouac site on almost every night: Figure 8.10).

In the statary phase raids are produced like the spokes of a wheel from the hub of the central bivouac. In all, including the first nomadic raid in and the last nomadic raid out, 15 raids radiate out from an average statary bivouac. On average these raids are 89 m long and 10 m wide, and this means that some overlap of foraging must occur near to the nest. However, the problem of this redundant foraging can be reduced because prey, such as roaches, spiders and other arthropods, quickly recover their normal density after a raid simply by migrating in from non-foraged areas. The army ants' social insect prey, of course, take much longer to recover. If the army ants could

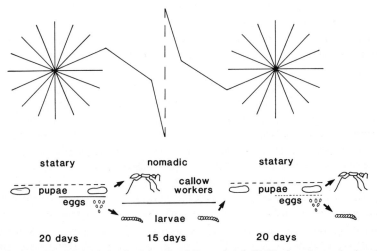

Figure 8.10 The 35-day behavioural cycle of *Eciton burchelli*. The foraging and migration pattern is diagrammatically represented above an outline of the brood cycle. For further explanation see text. (Redrawn from Franks and Fletcher, 1983.)

postpone sweeping over the same area for about a week they would encounter at least a significant proportion of their normal prey density. It is probably for this reason that army ants have a complicated spatial pattern of raiding in the statary phase. They separate successive raids by an average of 123° in a largely constant direction of rotation about the hub of the statary bivouac. This is a close approximation to the optimum angle of 126.4° that occurs in spiral leaf arrangement, as Leonardo da Vinci first recognized (Leigh, 1972). An angle of 126.4° results in neighbouring raids being separated by an average angle of 19.2°, close to the maximum possible with 15 radii, but this system conveys the special advantage of maximizing the separation of neighbouring foraging paths in time as well as space. Thus the army ants allow time for their general arthropod prey to recover before they are pillaged again. The angles between successive statary raids are far more variable than those between leaves in spiral phyllotaxis. However, this is unlikely to matter for two reasons: first, the army ant raids are analogous to extremely long thin leaves, being nine times as long as wide, so overlap is less likely; and second, since half the prey will have recovered any overlap is less important than complete shading of one leaf by another (Franks and Fletcher, 1983).

During the 15 days between successive statary phases the colony raids every day and usually emigrates at night down the principal trail of each day's raid system to a new bivouac site. However, on 14% of nomadic days the colony does not emigrate. On these occasions the ants return to the old bivouac site, striking out on the next day in a new direction on a new raid and later emigration. The average nomadic raid is 116 m long and the mean length of an emigration is 81 m. The problem of redundant foraging is probably even more serious during the nomadic phase than in the statary phase, because all the larvae are constantly requiring food and as they are all the same age they are equally prone to starvation. An army ant colony can reduce both the chance that it will cover its own tracks more than once within the same nomadic phase and also separate the areas swept out in successive statary phases by some form of navigation to straighten out its path over the whole of the nomadic phase. If the colony has no navigation system it would perform a simple random walk in the nomadic phase. In such a walk there would be on average 13 steps of 81 m each, allowing for days without emigrations. The average separation of the ends of such a random walk would be $\sqrt{13} \times 81$ m or 292 m. In fact the average separation of successive statary bivouacs is 529 m — significantly further than the army ants could achieve by a random walk. *E burchelli* colonies clearly navigate in the nomadic phase.

Franks and Fletcher (1983) analysed a series of alternative hypotheses to explain this navigation. Their conclusion was that *E. burchelli* colonies navigate by simply aligning each day's principal raid (and later emigration) with the average compass bearing of these activities on the previous day. This is probably achieved by the colony bivouacing some metres from the end of that day's raid and then beginning a new raid along the trail pheromone left from the previous day. This would help to give some continuity to the raid direction.

Further evidence that both the temporal and spatial pattern of raiding in *E. burchelli* is highly stereotyped comes from the single colony introduced to Orchid Island which maintained the same 35-day cycles of activity in which a spiral pattern of statary raiding alternated with linear nomadic navigation over the course of a year.

To determine the importance of the disturbance caused by *E. burchelli* colonies to the ecology of the leaf litter fauna as a whole it is important to know not only the search pattern of individual colonies but the entire search pattern of all the *E. burchelli* colonies on Barro Colorado Island. There are about 50 colonies of *E. burchelli* on the island and this number has remained constant for 50 years (Franks, 1982*a*). In total the raids of all these colonies will sweep over approximately 10×10^6 m^2 each year, an area equivalent to two-thirds of the island. But what is the overall pattern of attack? Because the foraging pattern of individual colonies is so predictable it is possible to assess the impact of the entire population of colonies. This has been accomplished through the construction of a computer simulation model. This model resembled a huge video game (Figure 8.11). The surface of the island was represented by a lattice of squares over which model colonies were programmed to raid in exactly the same temporal and spatial pattern as real colonies. Each square on the model island surface was programmed to increment its age by 1 each 'day' unless it had been raided, in which case its age returned to zero. In addition, each model colony kept a tally of all the ages of all the squares it raided. The equivalent density of colonies that occurs on the real island was used in the model, and the program was run for the equivalent of several 'model years' to reach equilibrium. After this steady state had been achieved, the model continued to run but now data was gathered on the condition of the model island and the conditions encountered by each colony. After one or more 'model' years, data could be gathered from the model island and from each model colony to determine the frequency distribution of the 'ages' of squares on the model forest floor, and the frequency at which each age class of square was raided by the army ants.

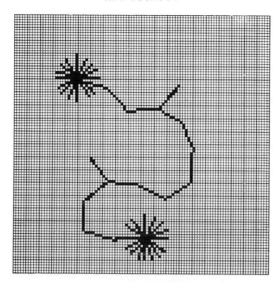

Figure 8.11 A scale drawing of the foraging pattern of an individual *E. burchelli* colony showing two statary phases and the intervening nomadic phase. Each small square in the arena represents a square of 10 m side. The figure represents the image on the VDU of a computer simulation of army ant raid patterns to determine how frequently areas are reraided. (From Franks and Bossert, 1983.)

As a precaution, the model also recorded the frequency at which the colonies collided with one another. The result was unexpected. In the model the colonies collided at a much higher frequency than could possibly be occurring on Barro Colorado Island. There are no reported observations of collisions between queen-right *E. burchelli* colonies, even though more than 1000 nomadic raids alone have been studied in Panama (Rettenmeyer, 1963; Willis, 1967; Schneirla, 1971). Thus real colonies must have some mechanism of mutual avoidance.

The only aspect of the foraging pattern of colonies that has yet to be explained was why colonies miss 14% of nomadic emigrations. Could it be that on these occasions colonies had encountered the pheromone trail of an alien colony which they could avoid by raiding and emigrating in a new direction? Franks and Bossert (1983) examined this possibility by programming each colony to leave its own marker in every square it visited and to avoid the squares that had been marked by other colonies within the last 20 days. Such avoidance was programmed as follows: when a colony came across a recently marked square it stopped raiding if it was in the statary phase and raided on another day. If it was in the nomadic phase it stopped raiding, also it did not emigrate but raided and emigrated in a new direction on the next day. With this refined behaviour the model colonies failed to emigrate on 14% of nomadic days and collided at such a low frequency that at this rate collisions would be so rare that they would go unobserved in the field. In effect the model colonies had been programmed to produce colony-specific trail pheromones that persisted for 20 days and created an aversive response like true territorial pheromones. Field observations suggest that these persistent trail pheromones are actually produced by *Eciton* colonies. Thus army ant colonies may also have spatio-temporal territories which produce spacing. This behaviour may help colonies to avoid areas that have been recently foraged by others. It may also prevent colonies coming into contact with one another in an accidental skirmish with a loss of part of their workforces. Avoidance behaviour could also be important because it might effect the frequency at which colonies return to areas previously raided by themselves.

When these behaviour patterns were incorporated, the simulation model suggested that *E. burchelli* colonies on Barro Colorado Island maintain the leaf litter fauna in a mosaic of patches of which a constant proportion, equal to half the island area, will have been raided within the last 240 days. Thus an almost equal area of the rainforest floor will be recovering from a raid and will be fully recovered. Purely by chance some areas will be raided so frequently that they will never recover, while others will be raided so rarely that they will resemble Orchid Island and also have a low diversity of ants in the leaf litter.

These army ants may therefore play a similar role to other so-called keystone predators (Paine, 1966) who remove dominant competitors and thereby provide space in which other species can flourish. By maintaining a patchwork of different community ages the army ants provide some space for each type of ant that exploits a different stage in the succession the army ants initiate. In this way army ants maintain diversity.

Thus the major influences on populations of army ants and their prey involve many of the same factors that are of prime importance in ant ecology in general; partitioning of space between colonies, searching efficiency during foraging, minimizing interference from other colonies, and the destruction of new colonies by established ones of the same species so that space is made available for other ant species.

The special features of ants, namely their sophisticated social organization, communication systems and structured divisions of labour, enable colonies to defend resources in their immediate neighbourhood and this often leads to extremely strong competition within a species. However, the very virulence of this intraspecific competition often puts a check on interspecific competition, and prevents competitive exclusion so that more ant species can occur in a habitat. The incredible species richness of ants in many habitats means that they interact with many other members of their communities and represent one of the major structural elements in almost all terrestrial ecosystems.

REFERENCES

Addicott, J. F. (1978a) The population dynamics of aphids on fire-weed; a comparison of local population and metapopulation. *Can. J. Zool.* **56**, 2554–2564.
Addicott, J. F. (1978b) Competition for mutualists. *Can. J. Zool.* **56**, 2093–2096.
Alexander, R. McN. (1982) *Locomotion of Animals*. Blackie, Glasgow and London.
Allies, A. B., Bourke, A. F. G. and Franks, N. R. (1986) Propaganda substances in the cuckoo ant *Leptothorax kutteri* and the slave-maker *Harpagoxenus sublaevis*. *J. Chem. Ecol.* **12**, 1285–1293.
Alloway, T. M. (1979) Raiding behaviour of two species of slave-making ants, *Harpagoxenus americanus* (Emery) and *Leptothorax duloticus* Wesson. *Anim. Behav.* **27**, 202–210.
Alloway, T. M. (1980) The origins of slavery in Leptothoracine ants. *Amer. Nat.* **115**, 247–261.
Amante, E. (1972) Preliminary observations on the swarming behaviour of the leaf cutting ant *Atta capiguara*. *J. Georgia entomol. Soc.* **7**, 82–83.
Anderson, M. (1984) The evolution of eusociality. *Ann. Rev. Ecol. Syst.* **15**, 165–189.
Anderson, R. E. (1970) An investigation of the colony odour of the ant *Lasius flavus*. Ph. D. Thesis, University of Cambridge.
Atsatt, P. R. (1981a) Ant-dependent food-plant selection by the mistletoe-butterfly *Ogyria amaryllis* (Lycaenidae). *Oecologia* **48**, 60–63.
Atsatt, P. R. (1981b) Lycaenid butterflies and ants; selection for enemy-free space. *Amer. Nat.* **118**, 638–655.
Ayre, G. L. and Blum, M. S. (1971) Attraction and alarm of ants (*Camponotus* spp.— Hymenoptera: Formicidae) by pheromones. *Physiol. Zool.* **44**, 77–83.
Bartz, S. H. and Hölldobler, B. (1982) Colony founding in *Myrmecocystus mimicus* Wheeler and the evolution of foundress associations. *Behav. Ecol. Sociobiol.* **10**, 137–147.
Beattie, A. J. (1985) *The Evolutionary Ecology of Ant-plant Mutualisms*. University Press, Cambridge.
Beattie, A. J. and Culver, D. C. (1981) The guild of myrmecochores in the herbaceous flora of West Virginia forests. *Ecology* **62**, 107–115.
Bentley, B. L. (1977) The protective function of ant visits to extrafloral nectaries of *Bixa orellana*. *Ecology* **65**(1), 27–38.
Bernstein, R. A. and Gobbel, M. (1979) Partitioning of space in communities of ants. *J. anim. Ecol.* **48**, 931–942.
Blum, M. S. (1981) *Chemical Defenses of Arthropods*. Academic Press, New York.
Blum, M. S. (1984) 'Alarm pheromones' in *Comprehensive Insect Physiology, Biochemistry and Pharmacology*, Vol. 9, *Behavior*, Kerkut, G. A. and Gilbert, L. I. (eds.) Pergamon Press, Oxford.
Blum, M. S., Jones, T. H., Hölldobler, B., Fales, H. M. and Jaouni, T. (1980) Alkaloidal venom mace: offensive use by a thief ant. *Naturwiss* **76**, 144–145.

Bonavita-Cougourdan, A. (1984) Activité antennaire et flux trophallactique chez la fourmi *Camponotus vagus* Scop. *Insectes sociaux* **30**(4), 423–442.
Bonavita-Cougourdan, A., LeMasne, G. and Rage, P. (1979) Une nouvelle méthode pour l'étude des échanges alimentaires chez les fourmis. *Insectes sociaux* **26**, 5–12.
Boosma, J. J., van der Lee, G. A. and van der Have, T. M. (1982) On the production ecology of *Lasius niger* in successive coastal dune valleys. *J. anim. Ecol.* **51**, 975–991.
Bosset, W. H. and Wilson, E. O. (1963) The analysis of olfactory communication among animals. *J. theoret. Biol.* **5**, 443–469.
Bourke, A. F. G. (1987) Worker reproduction in the higher eusocial Hymenoptera. In preparation.
Bradshaw, J. W. S., Baker, R. and Howse, P. E. (1975) Multicomponent alarm pheromones of the weaver ant. *Nature* **258**, 230–231.
Brandao, C. D. (1979) Division of labour within the worker caste of *Formica perpilosa* Wheeler. *Psyche* **85**, 229–237.
Brandt, D. C. (1980) Is the mound of *Formica polyctena* in origin a simulation of a rock? *Oecologia* **44**, 281–282.
Breen, J. (1977) Inter-nest battles of *Formica lugubris* in Ireland. *Entomol. Record* **89**, 296–298.
Brian, M. V. (1965) *Social Insect Populations*. Academic Press, London.
Brian, M. V. (1970) Communication between queens and larvae in the ant *Myrmica rubra*. *Anim. Behav.* **18**, 467–472.
Brian, M. V. (1973) Caste control through worker attack in the ant *Myrmica rubra*. *Insectes sociaux* **20**, 87–102.
Brian, M. V. (1974) Brood rearing in small cultures of the ant *Myrmica rubra*. *Anim. Behav.* **22**, 879–889.
Brian, M. V. (1975) Larval recognition by workers of the ant *Myrmica rubra*. *Anim. Behav.* **23**, 745–756.
Brian, M. V. (1977) *Ants*. Collins New Naturalist, London.
Brian, M. V. (1979) Caste differentiation and division of labour. In *Social Insects*, Hermann, H. R. ed. Vol. 1, Chapter 5. Academic Press, New York.
Brian, M. V. (1983) *Social insects: Ecology and Behavioural Biology*. Chapman and Hall, London.
Brian, M. V. and Abbott, A. (1977) The control of food flow in a society of the ant *Myrmica rubra*. *Anim. Behav.* **25**, 1047–1055.
Brian, M. V. and Brian, A. D. (1955) On the two forms macrogyna and microgyna of the ant *Myrmica rubra* L. *Evolution* **9**, 280–290.
Brian, M. V. and Downing, B. M. (1958) The nests of some British ants. *Proc. Xth Congr. Entomol.* **2**, 539–540.
Brian, M. V., Clarke, R. T. and Jones, R. M. (1981) A numerical model of an ant society. *J. anim. Ecol.* **50**, 387–405.
Brockman, H. J. (1984) The evolution of social behaviour in insects. In *Behavioural Ecology, an Evolutionary Approach* (2nd edn.), Krebs, J. R. and Davies, N. B. (eds.), Blackwell, Oxford, pp. 340–361.
Brown, J. H. and Davidson, D. W. (1977) Competition between seed-eating rodents and ants in desert ecosystems. *Science* **196**, 880–882.
Brown, J. H. and Orians, G. H. (1970) Spatial patterns in mobile animals. *Ann. Rev. Ecol. Syst.* **1**, 339–362.
Brown, J. L. (1964) The evolution of diversity in avian territorial systems. *Wilson Bull.* **76**, 160–169.
Brown, W. L. Jr (1955) The ant *Leptothorax muscorum* (Nyl.) in North America. *Entomol. News* **XX**, 43–50.
Buschinger, A. (1968) Mono- und Poly-gynie bei Arten der Gattung *Leptothorax* Mayr. *Insectes sociaux* **15**, 217–226.
Buschinger, A. (1970) Neue Vorstellungen zur Evolution des Sozialparasitismus und der Dulosis bei Ameisen. *Biol. Zentralbl.* **88**, 273–299.

REFERENCES

Buschinger, A. (1972) Kreuzung zweier Sozialparasitischer Ameisenarten, *Dorymyrmex pacis* Kutter und *Leptothorax kutteri* Buschinger. *Zool. Anz.* **189**, 169–179.

Buschinger, A. (1974a) Zur Biologie der sozialparasitischen Ameise *Leptothorax goesswaldi*. *Insectes sociaux* **21**, 133–144.

Buschinger, A. (1974b) 'Monogynie und Polygynie in Insektensozietaten.' In *Sozialpolymorphismus bei Insekten*, Schmidt, G. H. ed. Wissenschaftliche Verlagsgesellschaft, Stuttgart.

Buschinger, A. (1975) Sexual pheromones in ants. *Proc. Symp. Pheromones and Defensive Secretions of Social Insects*, Dijon 1974, 225–233.

Buschinger, A. (1976) Giftdrüsensekret als Sexualpheromon bei der Gastameise *Formicoxenus nitidulus*. *Insectes sociaux* **23**, 215–226.

Buschinger, A. (1981) Biological and systematic relationships of social parasitic Leptothoracini from Europe and N. America. In *Biosystematics of Social Insects*, P. E. Howse and J. L. Clement (eds.) Academic Press, New York.

Buschinger, A. (1985) New records of rare and parasitic ants in the French Alps. *Insectes sociaux* **32**, 3211–3324.

Buschinger, A., Ehrhardt, W. and Winter, U. (1980) The organisation of slave-raids in dulotic ants—a comparative study. *Zeits. Tierpsychol.* **53**, 245–264.

Buschinger, A. and Winter, U. (1983) Population studies of the dulotic ant *Epimyrma ravouxi* and the degenerate slave-maker *E. kraussi. Entomol. Gen.* **8**, 251–266.

Calabi, P., Traniello, J. F. A. and Werner, M. H. (1983) Age polyethism: its occurrence in the ant *Pheidole hortensis* and some general considerations. *Psyche* **90**, 395–412.

Cammaerts-Tricot, M. (1973) Phéromones agrégant les ouvrières de *Myrmica rubra. J. Ins. Physiol.* **19**, 1299–1315.

Carlin, N. F. (1982) Polymorphism and division of labour in the Dacetine ant *Orectognathus versicolor. Psyche* **88**, 231–244.

Carlin, N. F. and Hölldobler, B. (1983) Nestmate recognition in interspecific mixed colonies of ants. *Science* **222**, 1027–1029.

Carroll, R. and Janzen, D. H. (1973) Ecology of foraging by ants. *Ann. Rev. Ecol. Syst.* **4**, 231–257.

Chauvin, R. (1960) Facteurs d'asymetry et facteurs de régulation dans la construction du dôme chez *Formica rufa. Insectes sociaux* **8**, 201–205.

Cole, B. J. (1980) Repertoire convergence in two mangrove ants *Zacryptocerus varians* and *Componotus (Colobopsis) sp. Insectes sociaux* **27**, 265–275.

Cole, B. J. (1981) Dominance hierarchies in *Leptothorax* ants. *Science* **212**, 83–84.

Cole, B. J. (1983) Multiple mating and the evolution of social behaviour in the Hymenoptera. *Behav. Ecol. Sociobiol.* **12**, 191–201.

Cole, B. J. (1986) The social behaviour of *Leptothorax allardycei*: time budgets and the evolution of worker reproduction. *Behav. Ecol. Sociobiol.* **18**, 165–173.

Cottrell, C. B. (1984) Aphytophagy in butterflies: its relation to myrmecophily. *Zool. J. Linn. Soc.* **80**, 1–57.

Craig, R. and Crozier, R. H. (1979) Relatedness in the polygynous ant *Myrmecia pilosula. Evolution* **33**, 335–341.

Creighton, W. S. (1950) Ants of North America. *Bull. Mus. Comp. Zool. Harvard*, **104**.

Crewe, R. M. and Fletcher, D. J. C. (1974) Ponerine ant secretions: the mandibular gland secretion of *Paltothyreus tarsatus. J. entomol. Soc. S. Africa* **37**, 291–298.

Crozier, R. H. (1979) Genetics of sociality. In *Social Insects*, Vol. 1, H. R. Hermann (ed.) Academic Press, New York, 223–286.

Crozier, R. H., Pamilo, P. and Crozier, Y. (1984) Relatedness and microgeographic variation in *Rhytidoponera mayri*, an Australian arid zone ant. *Behav. Ecol. Sociobiol.* **15**, 143–150.

Culver, D. C. and Beattie, A. J. (1978) Myrmechory in *Viola. J. Ecol.* **66**, 53–72.

Darwin, C. (1859) *On the Origin of Species by Means of Natural Selection*. Murray, London.

Davidson, D. M. (1978a) Experimental tests of the optimal diet in two social insects. *Behav. Ecol. Sociobiol.* **4**, 35–41.

Davidson, D. M. (1978b) Size variability in the worker caste of a social insect (*Veromessor*

pergandei) as a function of the competitive environment. *Amer. Nat.* **112**, 523–532.
Davies, N. B. (1978) Ecological questions about territorial behaviour. In *Behavioural Ecology: an Evolutionary Approach*, Krebs, J. R. and Davies, N. B. (eds.) pp. 317–350. Blackwell Scientific Publications Oxford.
Davies, N. B. and Houston, A. I. (1984) Territorial economics in Krebs, J. R and Davies, N. B. *Behavioural Ecology: an Evolutionary Approach*, 2nd edn., pp. 148–169. Blackwell Scientific Publications, Oxford.
Dobrzanski, J. (1978) Problems of behavioural plasticity in the slave-making amazon ant *Polyergus rufescens* Latr. and in its slave ants *Formica fusca* L. and *F. cinerea* Mayr. *Acta Neurobiol. Exp.* **38**, 113–132.
Douglas, J. M. and Sudd, J. H. (1980) Behavioural coordination between ants and the aphids they attend. *Anim. Behav.* **28**, 1127–1139.
Duffield, R. M. and Blum, M. S. (1973) 4-Methyl-3-heptanone: Identification and function in *Neoponera villosa* (Hymenoptera; Formicidae). *Ann. ent. Soc. Amer.* **66**, 1357.
Duffield, R. M., Blum, M. S. and Wheeler, J. W. (1976) Alkyl-pyrazine alarm pheromones in primitive ants with small colonial units. *Comp. Biochem. Physiol.* **54B**, 439–440.
Dumpert, K. (1981) *The Social Biology of Ants*. Pitman, London.
Eickwort, G. C. (1981) Presocial Insects. In *Social Insects* Vol. 2, Hermann, H. R. (ed.) Academic Press, New York, 491.
Eisner, T. (1957) A comparative study of the proventriculus of ants. *Bull. Mus. comp. Zool. Harvard* **116**, 439–490.
Elmes, G. W. (1972) Observations on the density of queens in natural colonies of *Myrmica rubra* L. *J. anim. Ecol.* **42**, 761–771.
Elmes, G. W. (1974) The spatial distribution of a population of two ant species living in limestone grassland. *Pedobiologia* **14**, 412–418.
Elmes, G. W. (1976) Some observations on the microgyne form of *Myrmica rubra* L. *Insectes sociaux* **23**, 3–22.
Errard, C. (1984) Evolution en fonction de l'âge des relations sociales dans les colonies mixtes héterospecifiques chez les fourmis des genres *Camponotus* et *Pseudomyrmex*. *Insectes sociaux* **31**, 185–198.
Ettershank, G. (1971) Some aspects of the ecology and nest microclimatology of the meat-ant *Iridomyrmex purpureus*. *Proc. roy. Soc. Victoria* **84**, 137–152.
Evesham, E. J. M. (1984) Queen distribution movements and interactions in semi-natural nests of the ant *Myrmica*. *Insectes sociaux* **28**, 191–207.
Faber, W. (1967) Beitrage zur Kenntnis sozialparasitischer Ameisen 1. *Lasius (Austrolasius) reginae*, eine neue temporar sozialparasitischer Erdameise aus Oesterreich. *Pflanzenschutz Berichte* **36** (5–7), 73–107.
Fittkau, E. J. and Klinge, H. (1973) On biomass and trophic structure of the central Amazonian rain forest ecosystem. *Biotropica* **5**, 2–14.
Fletcher, D. J. C. and Blum, M. S. (1981) A bioassay technique for an inhibitory primer pheromone of the fire ant *Solenopsis invicta*. *J. Georgia entomol. Soc.* **16**, 352–356.
Forel, A. (1874) *Les fourmis de la Suisse*. Societé Helvétique des Sciences Naturelles, Zurich Imprimerie Cooperative, La Chaux de Fonde.
Franks, N. R. (1982*a*) A new method of censusing animal populations: the number of *Eciton burchelli* army ant colonies on Barro Colorado Island, Panama. *Oecologia* **52**, 266–268.
Franks, N. R. (1982*b*) Social insects in the aftermath of the swarm raid of the army ant *Eciton burchelli*. In *The Biology of Social Insects*, Breed, M. D., Michener, C. D. and Evans, H. E. (eds.) Westview Press, Boulder, Colorado.
Franks, N. R. (1982*c*) Ecology and population regulation in the army ant *Eciton burchelli*. In *The Ecology of a Tropical Forest*, Leigh, E. G. Jr, Rand, A. S. and Windsor, D. W. Smithsonian Institute Press, Washington, D. C. 389–395.
Franks, N. R. (1985) Reproduction, foraging efficiency and worker polymorphism in army ants. In *Experimental Behavioural Ecology: Fortschritte der Zoologie* BD 31, Hölldobler, B. and Lindauer, M. (eds.) G. Fischer-Verlag, Stuttgart.

REFERENCES

Franks, N. R. (1986) Teams in social insects: group retrieval of prey by army ants (*Eciton burchelli*), *Behav. Ecol. Sociobiol.* **18**, 425–429.
Franks, N. R. and Bossert, W. H. (1983) Swarm raiding army ants and the patchiness and diversity of a tropical leaf litter ant community. In *The Tropical Rain Forest*, Sutton, S. L., Chadwick, A. C. and Whitmore, T. C. (eds.) Blackwell, Oxford, 151–163.
Franks, N. R. and Fletcher, C. R. (1983) Spatial patterns in army ant foraging and migration: *Eciton burchelli* on Barro Colorado Island, Panama. *Behav. Ecol. Sociobiol.* **12**, 261–270.
Franks, N. R. and Hölldobler, B. (1987) Sexual competition during colony reproduction in army ants. *Biol. J. Linn. Soc.* in press.
Franks, N. R. and Norris, P. J. (1986) Constraints on the division of labour in ants: D'Arcy Thompson cartesian transformations applied to worker polymorphism. In *From Individual Characteristics to Collective Organisation, the Example of the Social Insects. Experientia (Supplement)* Pasteels, J. M. and Deneubourg, J. L. (eds.) in press.
Franks, N. R. and Scovell, E. (1983) Dominance and reproductive success among slave-making ants. *Nature* **304**, 724–725.
Free, J. B. (1956) A study of the stimuli which release the food-begging and offering responses of worker honeybees. *Brit. J. anim. Behaviour* **4**, 94–101.
Funk, R. S. (1975) Association of ants with ovipositing *Lycaena rubidus*. *J. Lepid. Soc.* **29**, 261–262.
Goodloe, L. and Sanwald, R. (1985) Host specificity in colony-founding by *Polyergus lucidus* queens. *Psyche* **92**, 297–302.
Gösswald, K. (1953) Histologische Untersuchungen an den arbeiterlosen Ameise *Teleutomyrmex schneideri* Kutter. *Mitt. Schweiz Entomol. Gesell.* **26**, 81–128.
Gösswald, K. (1954) *Unsere Ameisen*. Kosmos, Stuttgart.
Gotwald, W. H. Jr, (1982) Army ants. In *Social Insects*, H. R. Hermann (ed.), Vol. IV, Chapter 3. Academic Press, New York.
Greenberg, L. (1979) Genetic component of bee odour in kin recognition. *Science* **206**, 1095–1097.
Gris, G. and Cherix, D. (1977) Les grandes colonies de fourmis des bois du Jura (groupe *Formica rufa*). *Bull. Soc. ent. Suisse* **50**, 249–250.
Greenslade, P. J. M. (1979) *Ants of South Australia*. Special Educational Bulletin Series, South Australian Museum, Adelaide.
Hamilton, W. D. (1964) The genetical evolution of social behaviour I, II. *J. theoret. Biol.* **7**, 1–52.
Hangartner, W. (1967) Spezifität und Inaktivierung des Spurpheromons von *Lasius fuliginosus* und Orientierung der Arbeiterinnen im Duftfeld. *Z. vergl. Physiol.* **57**, 103–136.
Hangartner, W. (1969) Trail-laying in the subterranean ant *Acanthomyops interjectus*. *J. Insect Physiol.* **15**, 1–4.
Haynes, J. C. and Birch, M. C. (1984) In *Comprehensive Insect Physiology, Biochemistry and Pharmacology*, Vol. 9, *Behaviour*. Kerkut, G. A. and Gilbert, L. I. (eds.) Pergamon Press, Oxford.
Herbers, J. M. (1984) Queen-worker conflict and eusocial evolution in a polygynous ant species. *Evolution* **38**, 631–643.
Herbers, J. M. and Cunningham, M. (1983) Social organization in *Leptothorax longispinosus*. *Anim. Behav.* **31**, 775–791.
Hickman, J. C. (1974) Pollination by ants—a low energy system. *Science* **184**, 1290–1294.
Higashi, S. and Yamauchi, K. (1979) Influence of a supercolonial ant, *Formica* (*Formica*) *yessensis*, on the distribution of other ants in Ishikari coast. *Jap. J. Ecol.* **29**, 257–264.
Hölldobler, B. (1966) Futterverteilung durch Männchen im Ameisenstaat. *Zeit. vergl. Physiol.* **52**, 430–455.
Hölldobler, B. (1971) Recruitment behaviour in *Camponotus socius*. *Z. vergl. Physiol.* **75**, 123–142.
Hölldobler, B. (1973) Chemische Strategie beim Nahrungserwerb der Diebameise (*Solenopsis fugax*) und der Pharaoameise (*Monomorium pharaonis*). *Oecologia* **11**, 371–380.

Hölldobler, B. (1976a) Tournaments and slavery in a desert ant. *Science* **192**, 912–914.
Hölldobler, B. (1976b) Recruitment behaviour, home range orientation and territoriality in harvester ants, *Pogonomyrmex*. *Behav. Ecol. Sociobiol.* **1**, 3–44.
Hölldobler, B. (1977) The behavioural ecology of mating in harvesting ants. *Behav. Ecol. Sociobiol.* **1**, 405–423.
Hölldobler, B. (1978) Ethological aspects of chemical communication in ants. *Advances Study Behav.* **8**, 75–115.
Hölldobler, B. (1981) Foraging and spatio-temporal territories in the honey ant *Myrmecocystus mimicus*. *Behav. Ecol. Sociobiol.* **9**, 301–314.
Hölldobler, B. (1982) Interference strategy of *Iridomyrmex pruinosum* during foraging. *Oecologia* **52**, 208–213.
Hölldobler, B. (1983) Territorial behaviour in the green tree ant, *Oecophylla smaragdina*. *Biotropica* **15**, 241–250.
Hölldobler, B. and Carlin, N. F. (1985) Colony founding, queen dominance and oligogyny in the Australian meat ant *Iridomyrmex purpureus*. *Behav. Ecol. Sociobiol.* **18**, 45–58.
Hölldobler, B. and Haskins, C. P. (1977) Sexual calling in primitive ants. *Science* **195**, 793–794.
Hölldobler, B. and Lumsden, C. J. (1980) Territorial strategies in ants. *Science* **210**, 732–739.
Hölldobler, B. and Maschwitz, U. (1965) Der Hochzeitsschwarm der Rossameise *Camponotus herculeanus*. *Z. vergl. Physiol.* **50**, 551–568.
Hölldobler, B., Möglich, M. and Maschwitz, U. (1974) Communication by tandem running in the ant *Camponotus sericeus*. *J. Comp. Physiol.* **90**, 105–127.
Hölldobler, B., Stanton, R. C. and Markl, H. (1978) Recruitment and food retrieving behaviour in *Novomessor*. *Behav. Ecol. Sociobiol.* **4**, 163–181.
Hölldobler, B. and Wilson, E.O. (1977) The number of queens: an important trait in ant evolution. *Naturwiss.* **64**, 8–15.
Horstmann, K. (1974) Untersuchungen über dem Nahrungserwerb der Waldameisen: III Jahresbilanz. *Oecologia* **15**, 187–204.
Hubbard, M. D. (1974) Influence of nest material and colony odour on digging in the ant *Solenopsis invicta*. *J. Georgia Entomol. Soc.* **9**, 127–132.
Huber, P. (1810) *Recherches sur les Moeurs des Fourmis indigènes*. J. J. Paschoud, Paris.
Huxley, C. (1980) Symbioses between ants and epiphytes. *Biol. Rev.* **55**, 321–340.
Imamura, S. (1974) Observations on the hibernation of a polydomous ant, *Formica yessensis*. *J. Fac. Sci Hokkaido Univ.* **19** (series 2), 438–444.
Ito, M. (1973) Population trends and nest structures of *Formica yessensis*. *J. Fac. Sci Hokkaido Univ.* **19** (series 1), 270–293.
Jaisson, P. (1984) Social behaviour. In *Comprehensive Insect Physiology, Biochemistry and Pharmacology*, Vol. 9, *Behaviour*, Kerkut, G. A. and Gilbert, L. I. (eds.) Pergamon Press, Oxford.
Janzen, D. H. (1966) Coevolution of mutualism between ants and acacias in Central America. *Ecology* **20**, 249–275.
Janzen, D. H. (1968) Allelopathy by myrmecophytes: the ant *Azteca* as an allelopathic agent of *Cecropia*. *Ecology* **50**, 147–153.
Janzen, D. H. (1970) Herbivores and the number of tree species in tropical forest. *Amer. Nat.* **104**, 501–528.
Johnston, A. B. and Wilson, E. O. (1985) Correlates of variation in the major/minor ratios of the ant *Pheidole dentata*. *Ann. entomol. Soc. Amer.* **78**, 8–11.
Jutsum, A. R. (1979) Interspecific aggression in leaf-cutting ants. *Anim. Behav.* **27**, 833–838.
Kalmus, H. and Ribbands, C. R. (1952) The origin of the odour by which honeybees distinguish their companions. *Proc. roy. Soc.* (B) **140**, 50–59.
King, T. J. (1977) The plant ecology of ant hills on calcareous grasslands, *J. Ecol.* **65**, 235–278.
Kleinfeldt, S. E. (1978) Ant gardens: the interaction of *Codonanthe crassifolia* and *Crematogaster longispina*. *Ecology*, **59**, 449–456.
Krebs, J. R. (1978) Optimal foraging: decision rules for predators. In Krebs, J. R. and Davis,

N. B., *Behavioural Ecology: An Evolutionary Approach*, 2nd edn., Blackwell Scientific Publications, Oxford, 23–63.
Kugler, C. (1978) Evolution of the sting apparatus in the Myrmicine ants. *Evolution* **33**, 117–130.
Kutter, H. (1969) Die sozialparasitischen Ameisen der Schweiz. *Neujahrsblt. Naturforsch. Gesell. Zurich* **171**, 1–62.
Kwait, E. and Topoff, H. (1983) Emigration raids by slave-making ants: a rapid-transit system for colony relocations. *Psyche* **90**, 307–312.
Lamotte, M. (1947) Recherches écologiques sur le cycle saisonnier d'une savane guinéene. *Bull. Soc. Zool. France, Paris*, **72**, 88–90.
Lawton, J. H. and Heads, P. A. (1984) Bracken, ants and extrafloral nectaries: 1, components of the system. *J. anim. Ecol.* **53**, 995–1014.
Leigh, E. G. Jr (1972) The golden section and spiral leaf arrangement. In *Growth by Intussusception*, E. S. Deevey (ed.) *Trans. Connecticut Acad. Arts* **44**, 163–176.
Le Moli, F. and Mori, A. (1985) The influence of the early experience of worker ants on enslavement. *Anim. Behav.* **33**, 1384–1387.
Lenoir, A. (1972) Note sur le comportement de sollicitation chez les ouvrières de *Myrmica scabrinodis*. *C. R. Acad. Sci. Paris* D **274**, 705–707.
Lenoir, A. (1973) Les communications antennaires durent la trophallaxie entre ouvrières du genre *Myrmica*. *Proc VII Congr. IUSSI*, London 1973, 226–233.
Lenoir, A. (1979) Feeding behaviour in young societies of the ant *Tapinoma erraticum* L.: trophallaxis and polyethism. *Insectes sociaux* **26**, 19–37.
Levings, S. C. and Franks, N. R. (1982) Patterns of nest dispersion in a tropical ground ant community. *Ecology* **63**, 338–334.
Levings, S. C. and Traniello, J. F. A. (1981) Territoriality, nest dispersion and community structure in ants. *Psyche* **88**, 265–319.
Longhurst, C. and Howse, P. E. (1979) Some aspects of the biology of *Megaponera foetans*. *Insectes sociaux* **26**, 85–91.
Longhurst, C., Baker, R., Howse, P. E. and Speed, W. (1978) Alkylpyrazines in ponerine ants: their presence in three genera and caste specific behavioural responses to them in *Odontomachus troglodytes*. *J. Insect. Physiol.* **24**, 833–837.
Lyford, W. H. (1963) Importance of ants to brown podzolic soil genesis in New England. *Harvard Forest Paper* **7**, 1–18.
Mabelis, A. A. (1979) Wood ant wars: the relationship between aggression and predation in the red wood ant. *Netherlands J. Zool.* **29**, 451–620.
Mackay, W. P. and Mackay, E. E. (1984) Why do harvester ants store seeds in their nests? *Sociobiology* **9**, 31–47.
McNeil, J. N., Delisle, J. Finnegan, R. J. (1977) Inventory of aphids on 7 conifer species in association with the introduced Red Wood Ant. *Can. Entomol.* **109**, 1199–1202.
Malyshev, S. I. (1968) *The Genesis of the Hymenoptera and the Phases of their Evolution.* Methuen, London.
Markin, G. P. (1970) Food distribution within laboratory nests of the Argentine Ant, *Iridomyrmex humilis*. *Insectes sociaux* **17**, 127–158.
Marlin, J. C. (1968) Notes on a new method of colony foundation in *Polyergus lucidus lucidus*. *Trans. Illinois State Acad. Sc.* **61**, 207–209.
Maschwitz, U. and Hölldobler, B. (1970) Der Nestkartonbau bei *Lasius fuliginosus*. *Zeits. vergl. Physiol* **66**, 176–189.
Maschwitz, U. and Muhlenberg, M. (1973) *Camponotus rufoglaucus*, eine weglagernde Ameise. *Zool. Anz.* **191**, 364–368.
Maschwitz, U., Wüst, M. and Schurian, K. (1975) Bläulingsraupen als Zuckerlieferanten für Ameisen. *Oecologia* **18**, 17–21.
Michener, C. D. and Michener, M. H. (1951) *American Social Insects.* Van Nostrand Co., Toronto.

Mintzer, A. (1982) Nestmate recognition and incompatibility between colonies of the Acacia ant *Pseudomyrmex ferruginea*. *Behav. Ecol. Sociobiol.* **10**, 165–168.
Mirenda, J. T., Eakins, D. G., Gravelle, K. and Topoff, H. (1980) Predatory behaviour and prey selection by army ants in a desert-grassland habitat. *Behav. Ecol. Sociobiol.* **7**, 119–127.
Möglich, M. (1975) Recruitment of *Leptothorax*. *Proc. Symp. Pheromones Def. Secretions of Social Insects*, Dijon 1974, 235–242.
Möglich, M. and Alpert, G. D. (1979) Stone dropping by *Conomyrma bicolor*: a new technique of interference competition. *Behav. Ecol. Sociobiol.* **6**, 105–113.
Möglich, M. and Hölldobler, B. (1975) Communication and orientation during foraging and emigration in the ant *Formica fusca*. *J. Comp. Physiol.* **101**, 275–288.
Montagner, H. (1963) Etude préliminaire des relations entre les adultes et le couvain chez les guêpes sociales du genre *Vespa* au moyen d'un radio-isotope. *Insectes sociaux* **10**, 153–155.
Montagner, H. and Pain, J. (1973) Analyse du comportement trophallactique des jeunes abeilles. *C. R. Acad. Sci. Paris* D **272**, 297–300.
Muir, D. A. (1959) The ant–aphis–plant relationship in Dumbartonshire. *J. anim. Ecol* **28**, 133–140.
Naarman, H. (1963) Utersuchungen über Bildung and Weitergabe von Drüsensekreten bei *Formica* mit Hilfe der Radioisotopenmethode. *Experientia* **19**, 412–413.
Nielsen, M. G. and Jensen, T. F. (1975) Økologische studier over *Lasius alienus*. *Entomol. Medd.* **43**, 5–16.
Nonancs, P. (1986) Ant reproductive strategy and sex allocation theory. *Q. Rev. Biol.* **61**, 1–21.
O'Dowd, D. J. (1979) Foliar nectar production and ant activity on a neotropical tree *Ochroma pyramidalis*. *Oecologia* **43**, 233–248.
Olubajo, O., Duffield, R. M. and Wheeler, J. W. (1980) 4-Heptanone in the mandibular gland of the Nearctic ant, *Azcryptocerus varians*. *Ann. ent. Soc. Amer.* **73**, 93–94.
Orians, G. H. and Pearson, N. E. (1979) On the theory of central place foraging. In *Analysis of Ecological Systems*, Horn, D. J., Mitchell, R. and Stair, G. R. (eds.) Ohio State University Press, Columbus.
Oster, G. F. and Wilson, E. O. (1978) *Caste and Ecology in the Social Insects*. Princeton University Press, Princeton NJ.
Otto, D. (1958) Uber die Arbeitsteilung im Staate von *Formica rufa rufopratensis minor* Gösswald. *Wiss. Handl. Dtsch. Akad. Landwirtschaftwiss. Berlin* **30**, 1–169.
Paine, R. T. (1966) Food web complexity and species diversity. *Amer. Nat.* **100**, 65–75.
Pamilo, P. and Vario-Aho, S. L. (1979) Genetic structure of nests in the ant *Formica sanguinea*. *Behav. Ecol. Sociobiol.* **6**, 91–98.
Payne, T. L., Blum, M. S. and Duffield, R. M. (1975) Chemoreceptor responses of all castes of a carpenter-ant to male derived pheromones. *Ann. ent. Soc. Amer.* **68**, 385–386.
Pearson, B. (1981) The electrophoretic determination of *Myrmica rubra* microgynes as a social parasite and the possible significance of this with regard to the evolution of ant social parasites. In *Biosystematics of Social Insects*, Howse, P. E. and Clement, J. L. (eds.) Special volume, Systematics Society **19**.
Pearson, B. (1982) Relatedness of normal queens (macrogynes) in nests of the polygynous ant *Myrmica rubra*. *Evolution* **36**, 107–112.
Pearson, B. (1983) Intra-colonial relatedness amongst workers in a population of nests of the polygynous ant *Myrmica rubra*. *Behav. Ecol. Sociobiol.* **12**, 1–4.
Pearson, B. and Child, A. R. (1980) The distribution of an esterase polymorphism in macrogynes and microgynes of *Myrmica rubra*. *Evolution* **34**, 105–109.
Petal, J. (1978) The role of ants in ecosystems. In *Production Ecology of Ants and Termites*, M. V. Brian (ed.) Cambridge University Press.
Petralia, R. S. and Vinson, S. B. (1978) Feeding in the larvae of the imported fire-ant, *Solenopsis invicta*: behavioural and morphological adaptations. *Ann. ent. Soc. Amer.* **71(4)**, 643–648.

Pierce, N. E. (1984) Amplified species diversity: a case study of an Australian Lycaenid butterfly and its attendant ants. in Vane-Wright, R. I. and Ackery, P. R. (eds.); *Biology of Butterflies: 11th Sympt. roy. Ent. Soc. Lond.* Academic Press, London.

Pierce, N. E. and Mead, P. S. (1981) Parasitoids as selective agents in the symbiosis between Lycaenid butterfly larvae and ants. *Science* **211**, 1185–1187.

Pisarski, B. (1978) Comparison of various biomes. In *Production Ecology of Ants and Termites*, M. V. Brian. (ed.) Cambridge University Press.

Poldi, B. (1963) Studii sulla fondazione dei nidi nel Formicidae. *Sym. genet. biol. ital.* **12**, 132–199.

Pontin, J. A. (1960a) Field experiments on colony foundation by *Lasius niger* and *Lasius flavus. Insectes sociaux* **7**, 227–230.

Pontin, J. A. (1960b) Observations of the keeping of aphid eggs by ants of the genus *Lasius* in winter. *Entomol. mon. Mag.* **96**, 189–199.

Pontin, J. A. (1961) The prey of *Lasius niger* and *L. flavus. Entomol. mon. Mag.* **97**, 135–142.

Pontin, J. A. (1963) Further consideration of competition and the ecology of the ants *Lasius flavus* and *Lasius niger. J. anim. Ecol.* **32**, 565–574.

Pontin, A. J. (1978) The numbers and distribution of subterrranean aphids and their exploitation by the ant *Lasius flavus. Ecol. Entomol.* **3**, 203–207.

Poole, R. W. (1974) *An Introduction to Quantitative Ecology*. McGraw-Hill, New York.

Porter, S. D. and Jorgensen, C. D. (1981) Foragers of the harvester ant *Pogonomyrmex owyheei*: a disposable caste? *Behav. Ecol. Sociobiol.* **9**, 247–256.

Pudlo, R. J., Beattie, A. J. and Culver, D. C. (1980) Population consequences of changes in an ant-seed mutualism in *Sanguinaria canadensis. Oecologia* **46**, 32–37.

Regnier, F. E. and Wilson, E. O. (1971) Chemical communication and 'propaganda' in slave-maker ants. *Science* **172**, 267–269.

Rettenmeyer, C. W. (1963) Behavioural studies of army ants. *Univ. Kansas Sci. Bull.* **44**, 281–465.

Ross, K. G. and Fletcher, D. J. C. (1985) Comparative study of genetic and social structure in two forms of the fire ant *Solenopsis invicta. Behav. Ecol. Sociobiol.* **17**, 349–356.

Sabiti, J. M. N. (1980) The ecological role of ants in the ecosystem: foraging activity and excavation in *Paltothyreus tarsatus. Afr. J. Ecol.* **18**, 113–121.

Schemske, D. W. (1980) The evolutionary significance of extrafloral nectar production by *Costus woodsoni. J. Ecol.* **68**, 959–967.

Schmid-Hempel, P. (1984) Individually different foraging methods in the desert ant *Cataglyphis bicolor. Behav. Ecol. Sociobiol.* **14**, 231–271.

Schneirla, T. C. (ed.) H. R. Topoff (1971) *Army Ants, a Study in Social Organisation*. Freeman & Co., San Francisco.

Schneider, P. (1971) Vorkommen und Bau von Erdhügelnestern bei der Afghanischen Wüstenameise *Cataglyphis bicolor. Zool. Anzeig.* **187**, 202–213.

Smeeton, L. (1981) The source of males in *Myrmica rubra. Insectes sociaux* **28**, 263–278.

Snelling, R. R. (1981) A revision of the Honey Ants (Genus *Myrmecocystus*). *Sci. Bull. Nat. Hist. Mus. Los Angeles County*, **24**, pp. 23.

Soulié, J. (1961) Les nids et comportement nidificateur des fourmis du genre *Crematogaster. Insectes sociaux* **8**, 213–297.

Southwood, T. R. E. (1978) *Ecological Methods with Particular Reference to Insects*. (2nd edn.) Chapman & Hall, London.

Spangler, H. G. (1973). Vibration aids soil manipulation in Hymenoptera. *J. Kansas Entomol. Soc.* **46**, 157–160.

Stuart, R. J. (1984) Experiments on colony foundation in the slave-making ant *Harpagoxenus canadensis. Can. J. Zool.* **62**, 1995–2001.

Stuart, R. J. and Alloway, T. M. (1982) Territoriality and the origin of slave raiding in Leptothoracine ants. *Science* **215**, 1262–1263.

Stuart, R. J. and Alloway, T. M. (1983) The slave-making ant *Harpagoxenus canadensis* and its host species *Leptothorax muscorum*: slave raiding and territoriality. *Behaviour* **85**, 58–90.

Stumper, R, (1950) Les associations complexes des fourmis: commensalisme, symbiose et parasitisme. *Bull. Biol. Fr. Belge.* **84**, 376–399.
Sudd, J. H. (1957) Communication and recruitment in *Monomorium pharaonis*. *Anim. Behav.* **5**, 104–109.
Sudd, J. H. (1969) The excavation of soil by ants. *Zeits. Tierpsychol* **26**, 257–276.
Sudd, J. H. (1970) Specific patterns of excavation in isolated ants. *Insectes sociaux* **17**, 253–260.
Sudd, J. H. (1975) A model of digging behaviour and tunnel production in ants. *Insectes sociaux* **22**, 225–236.
Sudd, J. H. (1982) Ants: foraging, nesting, brood behaviour and polyethism. In *Social Insects* Hermann, H. R. (ed.), Academic Press, New York, pp. 108–155.
Sudd, J. H. (1987) Mutualism of ants and aphids. In *Aphids, their Biology, Natural Enemies and Control*, Harrewijn, P. and Minks, A. K. (eds.) Elsevier, Amsterdam.
Sudd, J. H. and Sudd, M. E. (1985) Seasonal changes in the response of wood-ants to sucrose baits. *Ecol. Entomol.* **10**, 89–97.
Sudd, J. H. and Sudd, M. E. (1987) The strategy of mixed diet in ants. In *From Individual Characteristics to Collective Organisation, the Example of the Social Insects.* Pasteels, J. M. and Deneubourg, J. L. (eds.) *Experientia (Supplement)* in press.
Szlep, R. (1970) The regulatory mechanism in mass foraging and recruitment of soldiers in *Pheidole*. *Insectes sociaux* **17**, 233–244.
Taylor, F. (1977) Foraging behaviour of ants: experiments with two species of Myrmicine ants. *Behav. Ecol. Sociobiol.* **2**, 147–167.
Taylor, R. W. (1978) *Nothomyrmecia macrops*: A living-fossil ant rediscovered. *Science* **201**, 979–985.
Thomas, J. A. (1984) The conservation of butterflies in temperate countries. In *Biology of Butterflies: 11th Symp. roy. Ent. Soc. Lond.* ed. Vane Wright & Ackery; Academic Press, London.
Topoff, H., Lamon, B., Goodloe, L. and Goldstein, M. (1984) Social and orientation behaviour of *Polyergus breviceps* during slave-making raids. *Behav. Ecol. Sociobiol.* **15**, 273–279.
Torossian, C. (1973) Etude des communications antennaires chez les Formicoidea. Analyse du comportement trophallactic des ouvrières de *Dolichoderus quadripunctatus* lors d'échanges alimentaires practiqués au bénéfice des sexués de la colonie. *C. R. Acad. Sci. Paris* D, **277**, 2073–2075.
Traniello, J. F. A. (1977) Recruitment behaviour, orientation and the organisation of foraging in the carpenter ant *Camponotus pennsylvanicus*. *Behav. Ecol. Sociobiol.* **2**, 61–79.
Traniello, J. F. A. (1978) Caste in a primitive ant; absence of age polyethism in *Amblyopone*. *Science* **202**, 770–772.
Traniello, J. F. A. (1980) Colony specificity in the trail pheromone of an ant. *Naturwiss.* **67**, 361–362.
Trivers, R. L. and Hare, H. (1976) Haplodiploidy and the evolution of the social insects. *Science* **191**, 249–263.
Tschinkel, W. and Bhatkar, A. P. (1974) Oriented mound-building in the ant *Trachymyrmex septentrionalis*. *Environ. Entomol.* **3**, 667–673.
Wallis, D. I. (1962) Aggressive behaviour in the ant *Formica fusca*. *Anim. Behav.* **10**, 267–274.
Walof, N. (1957) The effect of the number of queens of the ant *Lasius flavus* on their survival and on the rate of development of the first brood. *Insectes sociaux* **4**, 391–408.
Ward, P. S. (1983a) Genetic relatedness and colony organisation in a species complex of ponerine ants. 1. Phenotypic and genotypic composition of colonies. *Behav. Ecol. Sociobiol.* **12**, 285–299.
Ward, P. S. (1983b) Genetic relatedness and colony organisation in a species complex of ponerine ants. 2. Patterns of sex ratio investment. *Behav. Ecol. Sociobiol.* **12**, 301–307.
Watkins, J. F. II (1976) *The Identification and Distribution of New World Army Ants*. Baylor University Press, Waco, Texas.

Way, M. J. (1963) Mutualism of ants and Homoptera. *Ann. Rev. Entomol.* **8**, 307-344.
Weber, N. A. (1943) Parabiosis in Neotropical "ant gardens". *Ecology* **24**, 400-404.
Weber, N. A. (1982) Fungus ants. In *Social Insects*, Hermann, H. R. (ed.) Vol. IV, Chapter 4. Academic Press, New York.
Webster, R. P. and Nielsen, M. C. (1984) Myrmecophily in the Edward's Hairstreak (*Satyrium edwardsi*). *J. Lepid. Soc.* **38**, 124-133.
Wehner, R. (1970) Etudes sur la construction des cratères au-dessus des nids de la fourmi *Cataglypis bicolor*. *Insectes sociaux* **17**, 83-94.
Wesson, L. G. (1939) Contribution to the natural history of *Harpagoxenus americanus*. *Trans. Amer. entomol. Soc.* **65**, 97-122.
West-Eberhard, M. J. (1979) Sexual selection, social competition and evolution. *Proc. Amer. Phil. Soc.* **123**, 222-234.
West-Eberhard, M. J. (1981) Intra-group selection and the evolution of insect societies in *Natural Selection and Social Behaviour*. Alexander, R. D. and Tinkle, D. W. (eds.) Chiron, New York.
Wheeler, W. M. (1910) *Ants*. (reprinted 1960) Columbia University Press, New York.
Wheeler, W. M. (1916) Notes on some slave-raids of the western amazon ant *Polyergus breviceps*. *J. N. Y. entomol. Soc.* **24**, 107-118.
Willis, E. O. (1967) The behavior of bicolored ant-birds. *Univ. Calif. Publ. Zool.* **79**, 1-127.
Wilson, E. O. (1953) The origin and evolution of polymorphism in ants. *Q. Rev. Biol.* **28**, 136-156.
Wilson, E. O. (1958) A chemical releaser of alarm and digging behaviour in the ant *Pogonomyrmex badius*. *Psyche* **65**, 41-51.
Wilson, E. O. (1962) Chemical communication in the fire ant *Solenopsis saevissima* 1. The organisation of mass foraging; 2. An information analysis of the odour trail; 3. The experimental induction of social response. *Anim. Behav.* **10**, 134-164.
Wilson, E. O. (1965) Trail-sharing in ants. *Psyche* **72**, 2-7.
Wilson, E. O. (1971) *The Insect Societies*. Belknap Press of Harvard University Press, Cambridge, Mass.
Wilson, E. O. (1974) Aversive behaviour and competition within colonies of the ant *Leptothorax curvispinosus*. *Ann. entomol. Soc. Amer.* **67**, 777-780.
Wilson, E. O. (1975a) *Leptothorax duloticus* and the beginnings of slavery in ants. *Evolution* **29**, 108-119.
Wilson, E. O. (1975b) Enemy specification in the alarm-recruitment system of an ant. *Science* **190**, 798-800.
Wilson, E. O. (1975c) *Sociobiology: the Modern Synthesis*. Harvard University Press, Cambridge, Mass.
Wilson, E. O. (1976a) Behavioural discretization and the number of castes in an ant species. *Behav. Ecol. Sociobiol.* **1**, 141-154.
Wilson, E. O. (1976b) A social ethogram of the Neotropical arboreal ant *Zacryptocerus varians*. *Anim. Behav.* **24**, 354-363.
Wilson, E. O. (1978) Division of labor in fire ants based on physical castes. *J. Kansas entomol. Soc.* **51**, 615-636.
Wilson, E. O. (1980a) Caste and division of labor in leaf-cutter ants. I The overall pattern in *Atta sexdens*. *Behav. Ecol. Sociobiol.* **7**, 143-156.
Wilson, E. O. (1980b) Caste and division of labor in leaf-cutter ants. II The ergonomic optimization of leaf cutting. *Behav. Ecol. Sociobiol.* **7**, 157-165.
Wilson, E. O. (1983a) Caste and division of labor in leaf-cutter ants. III Ergonomic resilience in foraging by *A. cephalotes*. *Behav. Ecol. Sociobiol.* **14**, 47-54.
Wilson, E. O. (1983b) Caste and division of labor in leaf-cutter ants. IV Colony ontogeny of *A. cephalotes*. *Behav. Ecol. Sociobiol.* **14**, 55-60.
Wilson, E. O. (1984) The relation between caste ratios and division of labor in the ant genus *Pheidole*. *Behav. Ecol. Sociobiol.* **16**, 89-98.

Wilson, E. O. (1985a) The principles of caste evolution. In *Experimental Behavioural Ecology: Fortschritte der Zoologie Bd* **31**, Hölldobler, B. and Lindauer, M. (eds.) 307–324. G. Fischer, Stuttgart.

Wilson, E. O. (1985b) Between-caste aversion as a basis for division of labor in the ant *Pheidole pubiventris*. *Behav. Ecol. Sociobiol.* **17**, 35–37.

Wilson, E. O. and Fagen, R. M. (1974) On the estimation of total behavioural repertoires in ants, *J. N. Y. entomol. Soc.* **83**, 106–112.

Wilson, E. O. and Farish, D. J. (1973) Predatory behaviour in the ant-like wasp *Methocha stygia*. *Anim. Behav.* **21**, 292–298.

Wilson, E. O., Carpenter, F. M. and Brown, W. L. Jr. (1967) The first Mesozoic ants. *Science* **157**, 1038–1040.

Wisniewski, J. (1967) Die Zusammensetzung des Baumaterials der Nesthügel von *Formica polyctena* in Kiefernwaldern. *Waldhygiene* **7**, 117–121.

Wyatt, R. (1980) The impact of nectar robbing ants on the pollination system of *Asclepias curassavica*. *Bull. Torrey Bot. Club* **107**, 24–25.

Index

Acacia 33, 109, 123
Acanthomyops 38
Acanthomyops subterraneus 114
2-acetyl-3-methylcyclopentene 108
acidopore 52
Acromyrmex 36
active space 108, 114
Aculeata 25, 26
adaptive demography 75
Aenictus 32
aggression 101, 104
alarm pheromones 106, 150
　of aphids 129
alkyl sulphides 108
allelopathy 124
allometry 80–4, 89
allomone 139
Alloway 149, 157
altruism 12
amazon ant 142, 149
Amblyopone 23, 28, 32
Amblyopone pallipes 73, 74
Andersson 7
Anergates atratulus 141, 146–8
Aneuretus 37, 108
anholocyclic aphids 131
Anoecia 129
Anomma 32
antenna 25, 114
antennal movements in communication 100
Aphaeonogaster 34, 112, 125
　A. rudis 125
　A. subterranae 52
aphids 37, 127, 129
　alarm pheromones of 129
　eggs 123
　survival of 129
　subterranean 128, 131
Aphis fabae 129
　A. helianthi 129
　A. varians 129, 134
army ants 5, 46–50, 91–4, 117, 179–86

arrestant 52
Asclepias curassavica nectar 122
assemblage of males 118
Atta 41, 42, 87–91, 161, 165, 166, 172
　A. capiguara 117
　A. cephalotes 87, 90, 94, 95
　A. sexdens 87–9, 90
attendance of ants on aphids 128
Attini 35
attractant 108
Azteca 25, 108, 123
　A. chartifex 138

Bactris utilis 125
Baizongia pistaceae 131
Bartz and Hölldobler 9
bats 122
Beattie and Culver 161
behavioural elasticity, flexibility and plasticity 74, 94–6
behavioural repertoire 66–8, 84–6
Beltian bodies 123
Bernstein and Gobbel 164–5
biomass 121
bivouac 32
Bixa orelana, nectaries in 123
Boomsma *et al.* 168
Bothroponera tesserinodis 113
Brachyponera 32
Braconidae 132
bromeliads 129
brood 104
Brian 44, 162
Brian and Brian 145
Brown, J. 169
Brown, W. L. 161
Brown and Davidson 161
Brown and Orians 162
bull ants *see Myrmecia*
Buschinger 8, 9, 148, 157
Buschinger and Winter 158–9

Calabi *et al.* 73
callow workers 54, 103

INDEX

Camponotus 25, 102, 104, 180
　nests of 63
　C. beebi 138
　C. femoratus 138
　C. herculeanus 39, 53, 118
　C. ligniperdis 53
　C. pennsylvanicus 39, 108, 113
　C. rufoglaucus 137
　C. sericeiventris 67
　C. sericeus 113, 137
　C. (Colobopsis) sp. 68
　C. (Colobopsis) truncatus 79
Cardiocondyla 113
carpenter ants 39
carton nests 38, 62, 125
caste (including ratios) 67–73, 76–84, 86–97
Cataglyphis 39
　C. bicolor 59
　C. cursor 112
Cataulacus 35
Cecropia 37, 123
central place foraging 56, 162, 165, 171, 172, 176–8
Cerapachys 32
chemical communication 99
Cinara spp. 131, 134
citral 108
Cladosporium 63
claustral foundation 41
cocoon retrieval 104
Codonanthe crassifolia, nectaries in 125
coexistence 167
Cole 5, 19, 68
Colobopsis 39
colony budding 46
colony fission 46–9
colony foundation 40, 166
　claustral 41
colony growth
　ergonomic stage 41, 42
　founding stage 41
　reproductive stage 43
colony-odour 99
communication 110
　between ants and aphids 128
communities 161, 167, 179, 187
competition 155, 156, 162, 166–70, 179, 186
　exclusion 162, 186
　intraspecific 143, 166–8, 187
　between mutualists 133
　for space 162, 163, 165, 166, 186

concentration gradient of pheromone 109
conflicts 6, 7, 18, 22, 40, 75, 76
Conomyrma bicolor 164, 176
　C. insana 37
consumers, primary and secondary 121
convergent evolution 149, 153
cooperative hunting 112
Costus woodsoni, nectaries in 123
Craig and Crozier 12, 13
Crematogaster 35, 63
Crematogaster linata parabiotica 138
　C. longispina 125
crop 30
cuckoo ants 137, 146, 156

Dacetini 34
Daceton 34
Darwin 1, 75, 149, 154–6
Darwinian fitness 47
Davidson 178, 179
Davies 169
Davies and Houston 169
dietary breadth 179
division of labour 40, 65, 68, 69, 74–7, 84–9, 94, 160
deserts 34, 57, 60, 118, 124, 163, 172
diet and nest odour 101
diffusion role in chemical communication 99
4,6-dimethyl-4-octen-one 100
dimorphism, worker 34
dioecious aphids 131
discriminator, olfactory 100, 102
dispersal of seeds *see* myrmecochory
Dolichoderinae 28, 37, 36, 37
Dolichoderus quadripunctatus 51
dominance hierarchies 8, 18–23
Doronomyrmex pacis 145, 146
Dorylinae 28, 30, 32, 101
Dorylus 47, 170
Drepanosiphum pseudoplatanoides 129
Dufour's gland 53, 107, 108, 112, 114, 115, 118, 138, 141, 142, 150, 153

Eciton 32, 47–9, 56, 87, 186
Eciton burchelli 75, 79, 82, 83, 91–3, 179–85
　colony size in 48–9
　reproduction in 48–50
ecology 161–87
economics 40, 43, 47, 80, 84–91, 140, 160, 169
　of scale 41

Ectatomma 165
efficiency 159, 160
elaiosome 124
electrophoresis 6, 12–15
Elmes 17, 144–5
energy content of nectar 123
energy transformations 121
Emery's Rule 144, 156, 157
epicuticular lipids 103
Epilobium, aphids tended on 129
Epimyrma 149, 153
 E. kraussi 158, 159
 E. ravouxi 158, 159
 E. stumperi 141
epiphytes 125
ergonomics 41, 84–94, 160
eusociality 1, 5–7, 22, 24, 87
ethogram 66, 67, 77
Euxesta, nectaries in 123
Evesham 18
extra-floral nectaries 122

facultative myrmecophiles 129
fission, colony 46–7
Fittkau and Klinge 161
food exchange between workers 50
 antennal movements in 51
 donor 50, 53
 recipient in 50, 53
 signals in 51
foragers 53
foraging 91, 92, 162, 176, 177, 179, 182–3, 186
Forel 149
formic acid 107, 108
Formica 38, 107, 139, 149, 172
 F. aquilonia 38
 F. cunicularia 154
 F. fusca 38, 113, 142, 143, 149
 F. lemani 38, 143
 F. lugubris 39, 105, 128
  response to sugars in 134
 F. neoclara 38
 F. pallidefulva 38
 F. pergandei 150
 F. perpilosa 67
 F. polyctena 61, 104, 105
 F. pratensis 39
 F. rufa-group 34, 61, 105, 143
 F. rufibarbis 149
 F. sanguinae 12, 13, 38, 142, 151, 155, 156

F. subintegra 150, 153
F. yessensis 58
Formicinae 28, 38, 108, 142, 149, 150, 155, 181
Formicoidae 25
Formicoid complex 29
Formicoxenus 35, 118
Formicoxenus nitidulus 139
fostered brood 104
Franks 47–9, 81, 82, 91, 180, 181, 184
Franks and Bossert 180, 185
Franks and Fletcher 184
Franks and Hölldobler 49
Franks and Norris 82, 83
Franks and Scovell 20, 21
fungus 63

gland
 acid, in foragers 53
 Dufour's 53, 107, 108, 112, 114, 115, 118
 labial 63
 mandibular 106, 107
 metapleural 107
 poison 112, 118
 postpharyngeal 52
 pygidial 107
 radioactive tracers in 109
 sting 106, 113
 tergal 118
 venom 107
Glaucopsyche lygdamus 108
Gotwald 179
Gnamptogenus pleurodea 108
group recruitment 112, 113, 115, 118
guest ants 139

habituation 105
Hamilton 2, 3, 6
haplo-diploidy 2, 5, 22, 24, 117
haplometrosis 9, 19, 41
Harpagoxenus 140, 149, 158
 H. americanus 20, 21, 151–5, 159, 160
 H. canadensis 152, 155
 H. sublaevis 118, 142, 152, 153, 159
harvester ants 34, 40, 69, 124, 161, 165, 171, 178
Herbers 15
Herbers and Cunningham 77–78
(Z)-9-hexadecenal 113
hindgut 112

holocyclic aphids   131
Hölldobler   155, 165, 168, 175, 176
Hölldobler and Carlin   19
Hölldobler and Lumsden   168–74
Hölldobler and Wilson   19
Homoptera   37
honeybee   51
honeydew   52, 63, 127, 128
  sugars in   134
honey-pot ants   39, 172, 174–6
hostility, intercolonial   101, 105
Hydnophytum   125
Hymenoptera   24
Hypoponera opacior   107
Huber   149, 151

imprinting   104, 105, 153, 154
inbreeding   118
inclusive fitness   3, 4, 7, 10, 12, 15, 16, 40, 44, 45, 87, 95, 103, 169
  of myrmecophiles   134
information   98
infrastructure   55
inquilines   137, 140, 156
interference   165, 166, 175, 176
investment, schedule of   55
Iridiomyrmex   127
  I. humilis   17, 37, 114
  I. pruinosum   37, 164, 176
  I. purpureus   37, 19, 61
Isoptera   24

Janzen   167

kin selection   2, 7, 8, 12, 22

Large Blue Butterfly   133
larvae, recognition of   54
  sexual   54, 116
  attacked by workers   55
Lasioglossum zephyrum   101
Lasius   38, 142, 143
  L. alienus   143
  L. flavus   12, 38, 58, 59, 101, 103, 117, 128, 166–8
  L. fuliginosus   38, 63, 114, 143
  L. neoniger   172
  L. niger   53, 115, 143, 166–8
  L. reginae   143
  L. umbratus   143
leafcutter ants   35   (*see also Atta*)
learning in young workers   104
Leptogenys   32

Leptothorax   35, 75, 139–49, 153, 157
  L. acervorum   8, 63, 112, 118, 141, 142, 145, 146, 152
  L. allardycei   19, 20
  L. ambiguus   151, 155
  L. curvispinosus   18, 20, 66, 67, 155
  L. duloticus   149, 153, 155, 160
  L. gosswaldi   145–6
  L. gredleri   9, 152
  L. kutteri   8, 118, 141, 142, 145, 146, 153
  L. longispinosus   15, 20, 77, 78, 145, 151, 155
  L. muscorum   8, 152
  L. tuberum   8, 141
  L. unifasciatus   9, 158
  L. (Temnothorax) recedens   158
Levings and Franks   164, 180, 181
Levings and Traneillo   163
life-history strategies   159, 160
  of colony   43–9
Liometopum   38
Lycaena rubida   132
  population density in   129
Lycaenidae, aphytophagy of   131
  and Homoptera   133
Lysandra hispanica   132

macrogyne   17, 145
Macrosiphum valerianae   129
Maculinea arion   133
males   53, 107, 117
  of Dorylinae   33, 49
mandibles   25
mandibular gland   115, 118
Manica rubida   108
Maschwitz and Muhlenberg   137
mass recruitment   112, 113
mating   117, 118
meat ant   60
mechanical signals   99
Megaponera   32, 119
mellein   118
Melophorus   39
Membracidae   37
Messor   34
metapleural gland   25
Methoca   28, 57
methyl ketones   108
methyl 6-methyl-salicylate   118
2-methyl-4-heptanone   108
4-methyl-3-heptanone   108, 118
5-methyl-3-butyloctahydroindolizine   114

6-methyl-salicylate 108
6-methyl-5-heptenone 108
microgyne 17, 145
milking of aphids 128
mimicry, pheromonal 108
Mirenda 179
Monacis debilis 138
Moglich and Alpert 176
molecular weight of pheromones 99
monogyny 8, 14, 15, 146
Monomorium 34
M. pharaonis 17, 52, 105, 113, 114, 115, 118, 139
monomorphism 77, 79
motor display in recruitment 113, 115
Muller's organ of *Acacia* 123
mutualism 120
  of ants and insects 127
  and Homoptera 128
  and Lepidoptera 131
  and mimicry 136
  cost benefit, balance of 133, 134
  instability of 131
  competition between 133, 134
myrmecochory 34, 121
Myrmecia 12 – 14, 23, 28, 51
Myrmeciinae 31
Myrmecocystus 39, 59, 126, 149
  M. mimicus 10, 11, 155, 164, 174 – 6
Myrmecodia 125
Myrmecoid complex 28
myrmecophiles, population density of 129
myrmecophytes 121, 125
Myrmica 14, 23, 34, 43, 44, 46, 51, 139, 157
  M. hirsuta 144
  M. rubra 13, 17, 18, 52, 53, 54, 60, 108, 116, 144, 145
  M. ruginodis 144
  M. sabuleti 114, 144
  M. scabrinodis 133
Myrmicinae 28, 33, 107, 112, 149, 165, 181

nanitic workers 41
nearest neighbours 162
nectar composition 122
nectaries, extra-floral 122
Neivamyrmex 32
  N. nigrescens 179
non-independent colony foundation 140, 157

Neoponera villosa 107, 108
nest craters 57, 58
next mound 58, 60-1
nest odour 98
nests 55
  benefits of 56
  energetics of 58, 64
  excavation of 55, 57
  materials imported for 60
  specific differences in 57
  thermal properties of 61
nitrogen 36
nodes 29
noise in communication 100
nomadic phase of Dorylinae 32
Nothomyrmecia macrops 23, 28 – 9, 37, 107
Novomessor see Aphaenogaster
nurses 54

obligate myrmecophiles 129, 131
Ochroma pyramidalis, nectar in 122
3-octanol 108
3-octanone 5, 14, 26, 108
Odontomachus 32, 34, 107, 165, 181
Oecophylla 39, 56, 63 – 4, 69, 107, 108, 129, 170 – 2
Ogyris amaryllis 132
oligogyny 19
optimality 87, 90, 111, 160, 169, 176 – 9
Orians 176
oscillation 44
Oster and Wilson 42, 68, 77, 84
ovarioles 54

Pachycondyla 165, 181
Paine 186
Pamilo 13
Paltothyreus 32, 58, 108
Paraclitus cimiciformis 129
Paracryptocerus 79
parasitism 17, 137, 138, 140 – 8, 156, 160
parasites, temporary 142 – 3
parasitization rate of myrmecophiles 132
Paratrechina 181
  P. longicornis 17
parental care 7, 24, 65
parental manipulation 7
Pearson 13, 17, 145
Pharaoh's ant *see Monomorium pharaonis*

INDEX

Pheidole 34, 84, 86, 95 – 97, 116, 160, 164, 165, 179, 181
 P. dentata 76, 69 – 70, 73 – 74, 79, 86, 95
 P. guileimimuelleri 96
 P. pubiventris 97
pheromones 99, 106, 141 – 2, 144, 146, 150, 185, 186
 alarm 107, 150
 multiple 108
 physical properties of 99
 recruitment by 106, 111, 113 – 16
 territorial 106
phloem 127
physogastry 30, 32
Plagiolepis pygmaea 132
pleometrosis 9, 12, 41
Pogonomyrmex 34, 57, 61, 108 – 9, 118, 165, 171 – 2
 P. badius 109
 P. barbatus 173, 178
 P. occidentalis 178
 P. owyheei 69
 P. rugusos 173, 178
poison gland 112, 118, 139, 146
pollination 122
polydomous ants 15, 56, 151
Polyergus 38, 142, 149 – 51, 155, 160
 P. breviceps 142, 149
 P. lucidus 142, 149, 155
 P. rufescens 149, 155
Polygonum cascadense 122
polygyny 5, 8 – 12, 19, 23, 145, 157
Polyrachis 39
 P. simplex 62
Ponera 32
 P. eduardi 113
 P. pennsylvanica 107
Ponerinae 32, 51, 165
poneroid complex 27, 29
Pontin 166, 168
populations of atended aphids 129
post-petiole 28
predation 106, 120
 of aphids by ants 129, 134
primary consumers 121, 128
profitability of prey 176 – 9
Proformica 39
propaganda substances 139, 142, 150, 157
proventriculus 30, 38
Pseudomyrmex 33, 123
 P. feruginea 101, 103, 105
 P. triplaridis 37
Pseudomyrmicinae 2, 9, 20, 23

Pteridium aquilinum nectaries 123, 134
Pterochloroides persicae 129
Puss Moth 107
pyrazines 107

queen 53 – 4, 103, 116
queen control 144
queen substance 103, 116

Raptiformica 38
recognition of larvae 54
 of nestmates 98, 100 – 105
  genetic basis of 105
  signals in 99
recovery in foraging 111
recruitment 111, 171
recruitment pheromones 106, 112
recruitment trails 138, 149, 157
regurgitation 30, 50 – 2, 105, 117
relatedness 3 – 6, 13, 105, 117
relationship, coefficient of 3, 102
repellent chemicals 107
reproduction 75
 of colony 43, 46
 strategy of 43
resource flow 45, 50 – 5
 in society 53
 to larvae 51, 54
 to queen 51
resource sink 53 – 4
Rhytidoponera 15, 17, 32, 118
ritualization 175
role 65, 73
Rossomyrmex 149

Sanguinaria canadensis 124
scale, economies of 41
Scleroderma 26
search in foraging 111
seed-bank 124
seed transport 124
Serviformica 38
sex-determination 24
sex pheromones 108, 116 – 18
sex ratios 4, 6, 14
sexuals 106
signals 98
silk nests 119
siphuncles 129
slave-making ants 104, 137, 140, 142, 149 – 60
 degeneracy of 158
 evolution of 154

slave-making, intraspecific 156, 173
Smithistruma 34
social homeostasis 94
Solenopsis 87, 165
  S. (Diplorhoptrum) fugax 35, 138
  S. geminata 67, 79, 84, 178
  S. invicta 15, 35, 54, 57, 79, 84, 113 – 15
soil formation, ants and 58
solicitation 50 – 1
solid foods 50, 52
spatial patterns (*see also* nearest neighbours) 162, 163 – 5, 167
spatio-temporal territorialism 174
specialization 53
speciation 144, 157
Sphecomyrmma freyi 26 – 8
statary phase of Dorylinae 32
sting gland 107, 113
stridulation 50, 57, 100
Strongylognathus 149, 158
stylets of aphids 127
subcastes 34
succession 181, 186
super-colony (*see also* unicolonialism) 105
switches, developmental 45
Symydobius oblongus 128
Syrphidae 127

Tachinidae 132
tandem calling 111
tandem running 152
Tapinoma 108
  T. erraticum 52
task 66, 71, 79
task fixation 74
teams in nest building 63
Teleutomyrmex schneideri 141, 146
temporal polyethism 69 – 75
tergal gland 118
termites 24

territorial conflict 106
territorial pheromones 107
territoriality 154, 157, 168 – 76
Tetramorium 34, 146, 148
  T. caespitum 52, 141, 146, 156
Tetraneura ulmi 131
thief ants 138
time budgets 77 – 8
Tiphioidea 26, 57
Trachymyrmex 36, 59
trails 113 – 16
trail pheromone 114, 185, 186
trehalose 132
Triplaris americana 37
Trivers and Hare 4, 6
trophallaxis 30, 50 – 2
trophic eggs 52
trophic level 121
tubulation of gaster 28 – 30, 33

undecane 108
unicolonialism (*see also* super-colony) 16

Veromessor 34, 164
  V. pergandei 179
visa, olfactory 100, 102, 104, 141, 154

waggle display 111, 115
wars between colonies 105
wasps 51, 127
weaver ant *see Oecophylla*
West-Eberhard 76
Wheeler 141, 156, 158
Wilson 18, 66 – 7, 70, 79, 84, 89, 114 – 15, 154
wing-casting 116
wood-ants *see Formica rufa* group

Zacryptocerus 35, 79, 107
  Z. varians 67 – 68
(Z)-9-hexadecenal 114